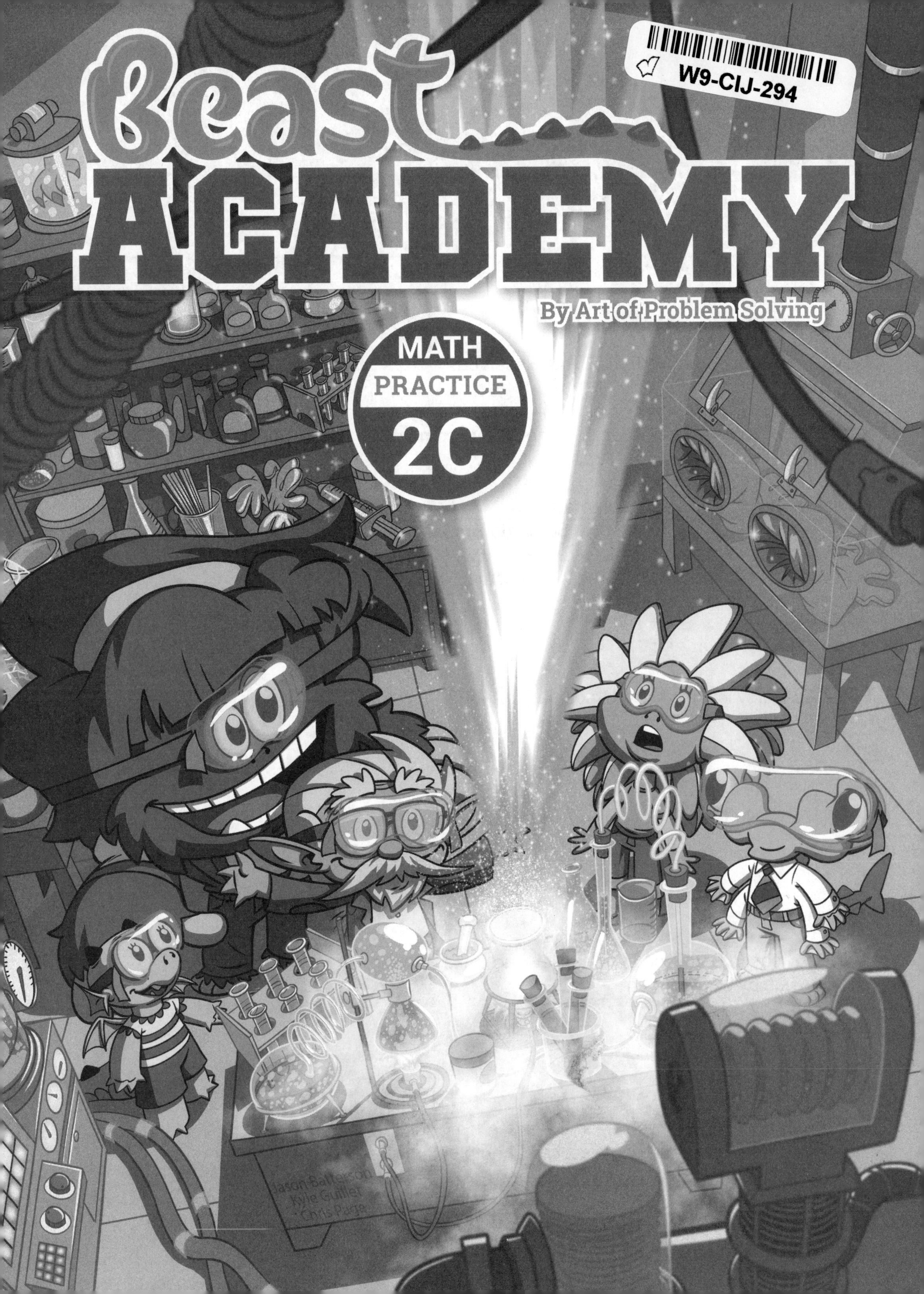
W9-CIJ-294
Beast
ACADEMY
By Art of Problem Solving
MATH
PRACTICE
2C
Jason Batterson
Kyle Guillet
Chris Page

Published by: AoPS Incorporated
10865 Rancho Bernardo Rd Ste 100
San Diego, CA 92127-2102
info@BeastAcademy.com

ISBN: 978-1-934124-35-2

Written by Jason Batterson, Kyle Guillet, and Chris Page
Book Design by Lisa T. Phan
Illustrations by Erich Owen
Grayscales by Greta Selman

Visit the Beast Academy website at BeastAcademy.com.
Visit the Art of Problem Solving website at artofproblemsolving.com.
Printed in the United States of America.
2021 Printing.

Contents:

This is Practice Book 2C in a four-book series.

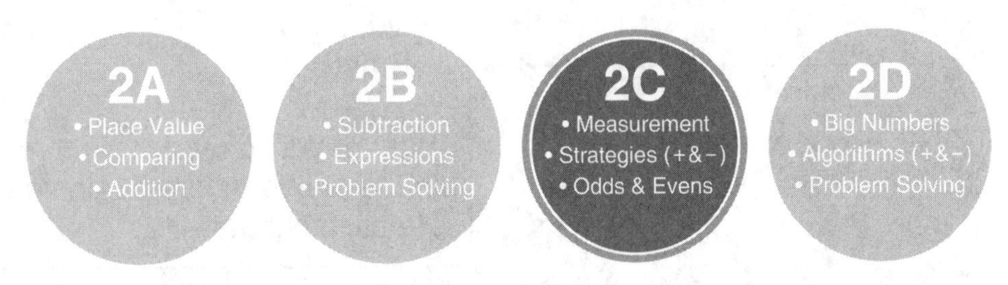

For more resources and information, visit BeastAcademy.com.

INTRODUCTION

How to Use This Book

This is Beast Academy Practice Book 2C.

Each chapter of this Practice book corresponds to a chapter from Beast Academy Guide 2C.

MATH PRACTICE 2C

The first page of each chapter includes a recommended sequence for the Guide and Practice books.

You may also read the entire chapter in the Guide before beginning the Practice chapter.

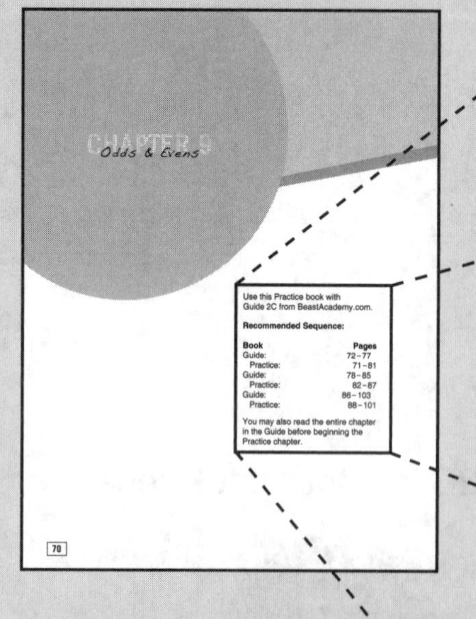

Use this Practice book with Guide 2C from BeastAcademy.com.

Recommended Sequence:

Book	**Pages**
Guide:	72–77
Practice:	71–81
Guide:	78–85
Practice:	82–87
Guide:	86–103
Practice:	88–101

You may also read the entire chapter in the Guide before beginning the Practice chapter.

Some problems in this book are very challenging. These problems are marked with a ★. The hardest problems have two stars!

Every problem marked with a ★ has a ***hint!***

Hints for the starred problems begin on page 102.

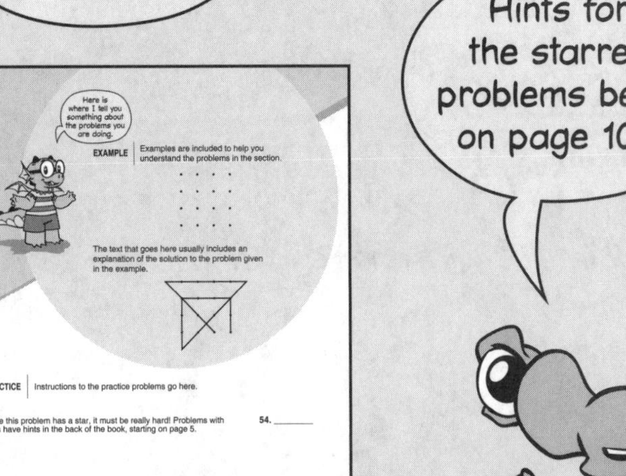

PRACTICE | Instructions to the practice problems go here.

54. ★ Since this problem has a star, it must be really hard! Problems with stars have hints in the back of the book, starting on page 5. 54. ______

55. After problem 54 is problem 55. All of the problems in each chapter are numbered in order, starting with problem number 1.

42 Guide Pages: 39-43

54. ★

42 Guide Pages: 39-43

Some pages direct you to related pages from the Guide.

None of the problems in this book require the use of a calculator.

Solutions are in the back, starting on page 106.

A complete explanation is given for every problem!

Pretend Solutions

54. Each of the solutions in the back of this book includes a complete explanation of how to solve the problem, not just the answer. Sometimes, we will provide a solution that is different than the way you solved the problem. It is useful to read the solution to every problem even if you got the answers correct. You may learn a different way to solve the problem. Or, you may find a mistake that we made! Wouldn't it be fun to write the authors to point out a mistake they made.

55. These are just fake solutions that we wrote so that it looked like there was text here. If we put the real solutions here, it would give away some of the answers to the problems that show up in the book. Then, you wouldn't be able to have fun trying to solve them on your own! Here are some fake diagrams to go with the fake solutions:

56. If you are having trouble with a problem, it can be tempting to just look in the back of the book for the answer. The more you think about a problem, the more you learn by solving it (this is usually true even if you get the wrong answer). Do your best to figure out a problem before looking for the answer. Don't forget to use the hints for the starred problems before reading the solution.

57. Here are some more diagrams to make these fake solutions look like real ones:

58. Here is some mathy stuff:

$51+52+53=50+50+50+1+2+3=150+6=156.$

So, the answer is **156**. Notice that the answer is bold. This makes it easier to find the answer to each problem quickly.

59. The header below separates sections of problems. This makes it easier to find the solution you are looking for.

Fake Solutions

60. Did you notice the number on the right side of the header above? That is the page number where you can find the problems in the Practice book. That way, when you are all finished reading a solution, you can flip back to the page you were working on.

61. You may be asking yourself, "Why did the author go to so much trouble writing text that no one can read without super-human vision?" There's no good answer, really.

62. Sometimes we give two possible ways to solve a problem. Usually, these two ways are separated by the word "or" and some lines like the ones below:

— or —

The second solution would come here. Generally, the quicker or more clever solution comes second.

63. Here is another fake solution with some diagrams. It's from a different book in the Beast Academy series. Can you figure out what is going on just by looking at the diagrams?

12, 8, 3, 8, 15, 8

Below is a new diagram with the same pieces.

12, 3, 15, 8

Can you guess what the question that goes with the diagrams above was? Here's a hint: The answer is **240 squares**.

64. There's another header below.

Complete Jibberish

65. The jibberish below is called Lorem Ipsum. It is complete jibberish used as placeholder text to fill the rest of the space below.

66. Curabitur velit nisl, eleifend nec interdum id, commodo id magna. Nam venenatis neque ac augue ultrices volutpat. Aenean quis diam ut ligula volutpat dapibus.

67. Makes no sense does it? I think some of those words are Latin.

68. Morbi rutrum consequat eros a rutrum. Praesent pellentesque urna ac nibh dictum consequat. Maecenas eget velit magna. Duis volutpat auctor neque sit amet congue. Aliquam erat volutpat. Suspendisse potenti. Pellentesque ut nulla lorem. Morbi consectetur purus volutpat enim vestibulum rhoncus sodales ligula aliquet.

69. Ready to get started on some math problems? Get to it!

Pretend Solutions 111

CHAPTER 7 Measurement

Use this Practice book with Guide 2C from BeastAcademy.com.

Recommended Sequence:

Book	**Pages:**
Guide:	14-20
Practice:	7-11
Guide:	21-33
Practice:	12-25
Guide:	34-41
Practice:	26-35

You may also read the entire chapter in the Guide before beginning the Practice chapter.

EXAMPLE

What coins are larger than the circle below?
What coins are smaller than the circle below?

The circle above is the same size as a U.S. nickel.

U.S. pennies and dimes are smaller than the circle. U.S. quarters, half-dollars, and dollar coins are larger than the circle.

You can compare coins from other countries to the circle as well.

PRACTICE

Circle the correct answer to each question below. You can find printable outlines of each item at BeastAcademy.com.

1. Is the line below longer or shorter than the long edge of a standard playing card (for example, the Ace of Spades)?

1. longer shorter

2. Is the line below longer or shorter than a AA battery?

2. longer shorter

3. Is the line below longer or shorter than the long edge of a U.S. dollar bill?

3. longer shorter

4. Is the long edge of a standard sheet of printer paper longer or shorter than the long edge of a page in this book?

4. longer shorter

EXAMPLE | Which of the two lines below is longer?

We can use a small object to compare the two lines. Below, we use a pen cap. The line on the left is slightly longer than the top of the pen cap. The line on the right is slightly shorter than the top of the same pen cap.

So, the **line on the left** is just a little bit longer than the line on the right.

PRACTICE | Use an object to help you compare each pair of items below.

5. Circle the longer crayon.

6. Circle the longer matchstick.

7. Circle the longer paper clip.

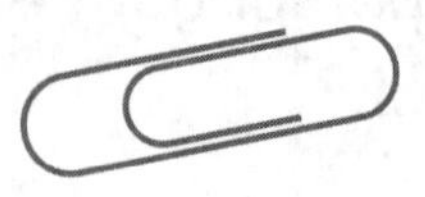
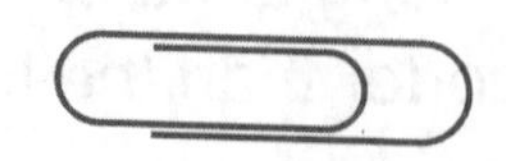

You can also compare items by marking their lengths on a piece of paper.

We see from our marks that the key on the right is longer than the key on the left.

PRACTICE | Label each set of items below from shortest to longest.

8. Number the lines below from shortest (1) to longest (3).

9. Number the fish below from shortest (1) to longest (3).

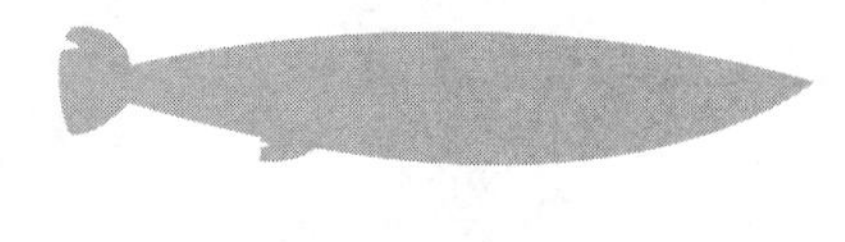
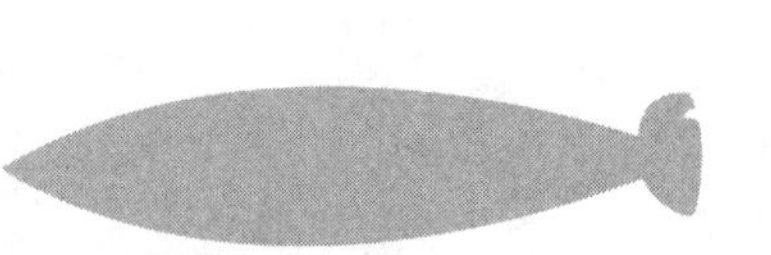
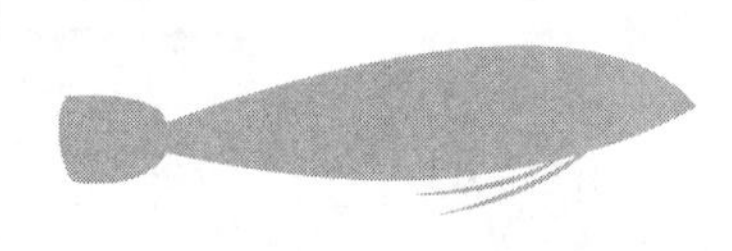

10. Number the shoes below from shortest (1) to longest (3).

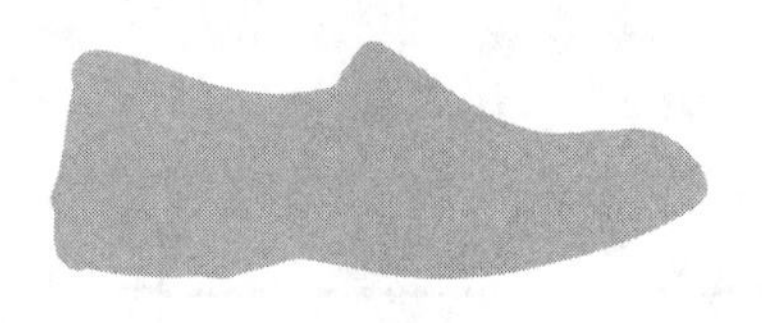

To compare items by measuring, we have to use the same ***unit of measurement***.

A ***unit of measurement*** is anything that can be used to measure.

You can use just about anything as a unit of measurement. For example, we can use the distance across a dime as a unit of measurement.

You can place four dimes on the circles above to see that the line is 4 dimes long.

PRACTICE | Write the length of each line below in dimes.
You can find a printable dime outline at BeastAcademy.com.

11. ______dimes

12. ______dimes

13. ______dimes

14. ______dimes

15. ______dimes

A ***ruler*** is a tool used to measure short lengths and distances. You will learn to use a standard ruler on the next few pages to measure items in inches and centimeters. But first, make your own ruler with your very own unit of measurement!

Step 1:

Get a sheet of paper and fold it in half three times, as shown below. You can use a little tape to hold the last fold.

Step 2:

Mark your ruler. Use your thumb to make equally-spaced marks on your ruler, starting at one end. Label these marks, starting with 0 at one end, and ending when you run out of room.

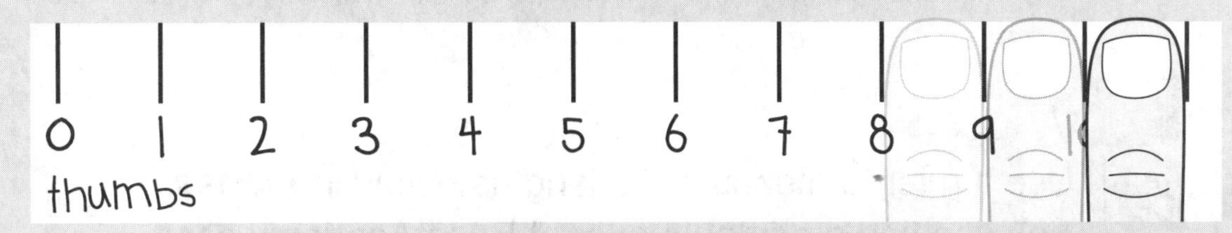

Depending on the size of your thumb, you may have a different number of thumbs on your ruler than the one above.

Step 3:

Start measuring! How many thumbs long is your pencil? A fork? The TV remote? A tube of toothpaste? Your pet hamster? A ham sandwich?

How many thumbs long is a banana? Can you make a longer ruler to measure items in bananas? How many bananas long is your bed?

You can measure items in bananas and thumbs! Then, you can measure your height using both bananas and thumbs and say silly stuff like, "Last year, I was only 5 bananas and 9 thumbs tall, but I grew 5 thumbs!"

Most rulers have measurements on both sides.
One side is used for measuring length in ***inches*** and the other is used for measuring length in ***centimeters***.

Inches and centimeters are units of length. We use "in" and "cm" as abbreviations for inches and centimeters.

To measure an object, place the line marked 0 on the ruler at one end of the object. The number at the other end of the object is its length.

On this page, we measure objects using inches.
The side of a ruler labeled "in" is marked in inches.

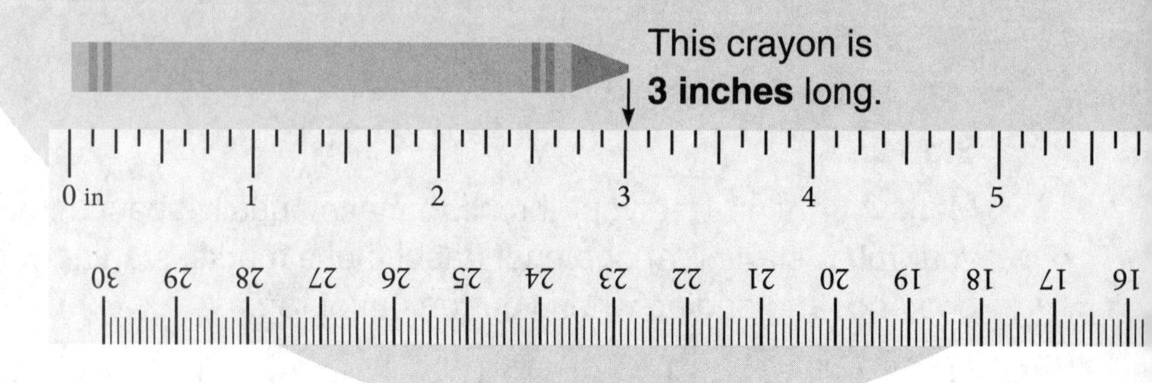

PRACTICE | Use a ruler to measure the lengths below in ***inches***.
You can find a printable ruler at BeastAcademy.com.

16. How many inches long is the line below? **16.** ______ in

17. How many inches long is the line below? **17.** ______ in

18. How many inches long is the line below? **18.** ______ in

On this page, we measure objects in centimeters. The side of a ruler labeled "cm" is marked in centimeters.

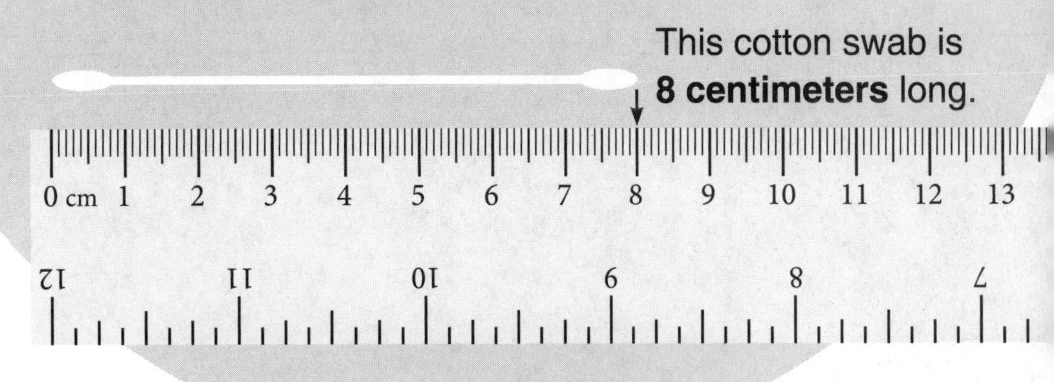

PRACTICE | Use a ruler to measure the lengths below in ***centimeters***.

19. How many centimeters long is the line below? **19.** _______ cm

20. How many centimeters long is the line below? **20.** _______ cm

21. Label the length of each side of the triangle below in centimeters.

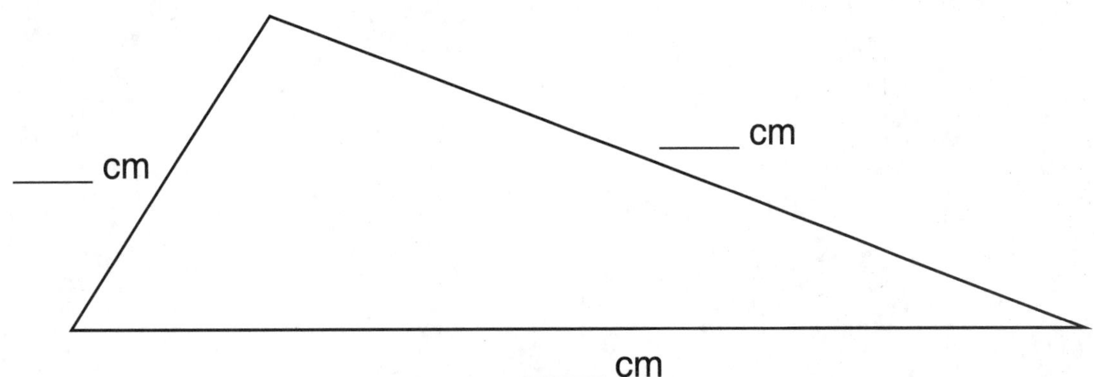

PRACTICE | Use a ruler to help you complete each task below.

22. Draw a straight line that connects the two dots above that are 2 inches apart. Write "2 inches" on this line.

23. Draw a straight line that connects the two dots above that are 3 inches apart. Write "3 inches" on this line.

24. Draw a straight line that connects the two dots above that are 4 inches apart. Write "4 inches" on this line.

25. Connect three of the dots below to create a triangle with sides that are 2 inches, 3 inches, and 4 inches long.

PRACTICE | Use a ruler to help you complete each task below.

26. Draw a dot on the line below that is 10 centimeters from the dot marked F.

F

27. Draw a dot on the line below that is 11 centimeters from the dot marked G.

G

28. Draw a dot on the line below that is 12 centimeters from the dot marked H.

H

29. Draw ***two*** dots on the line below that are each 5 centimeters from dot J.

J

In a **Measure-Maze**, the goal is to trace a path along the given lines that connects all of the points in the maze.

The distances between the dots on the path must match the given distances in order.

EXAMPLE | Complete the following Measure-Maze using the distances 2 cm, 3 cm, 3 cm, and 7 cm in order.

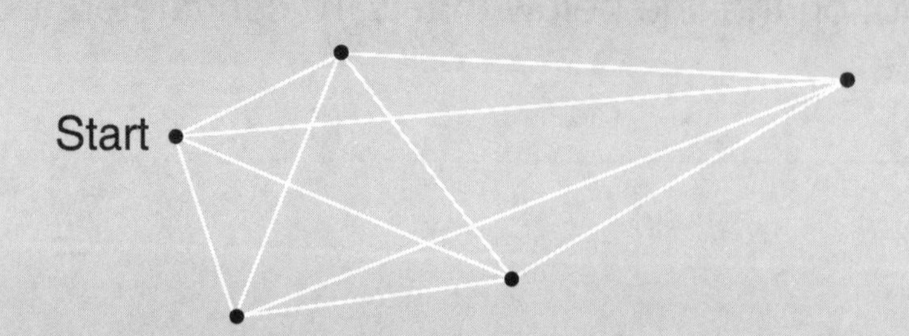

The only way to connect all of the dots using the given distances in order is shown below.

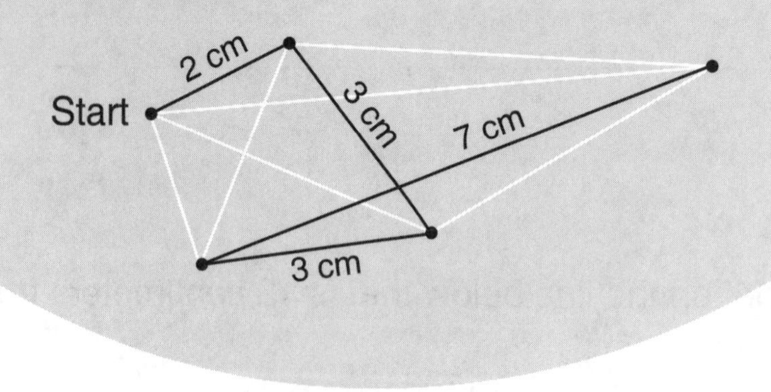

PRACTICE | Solve each Measure-Maze below.

30. **Distances:**
1 cm, 3 cm, 4 cm.

Start

31. **Distances:**
4 cm, 5 cm, 1 cm.

Start

PRACTICE | Solve each Measure-Maze below.

32. **Distances:**
3 cm, 5 cm, 4 cm.

Start

33. **Distances:**
1 cm, 2 cm, 2 cm, 5 cm.

Start

34. **Distances:**
4 cm, 1 cm, 2 cm, 3 cm.

Start

35. **Distances:**
4 cm, 4 cm, 1 cm, 2 cm.

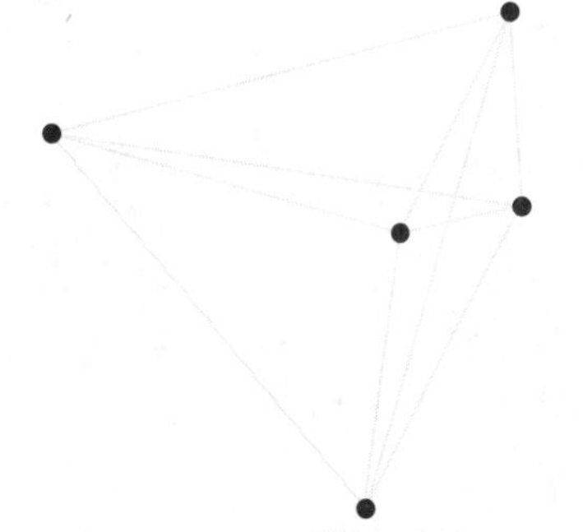

Start

PRACTICE | Solve each Measure-Maze below.

36. **Distances:**
2 cm, 1 cm, 5 cm, 1 cm, 5 cm.

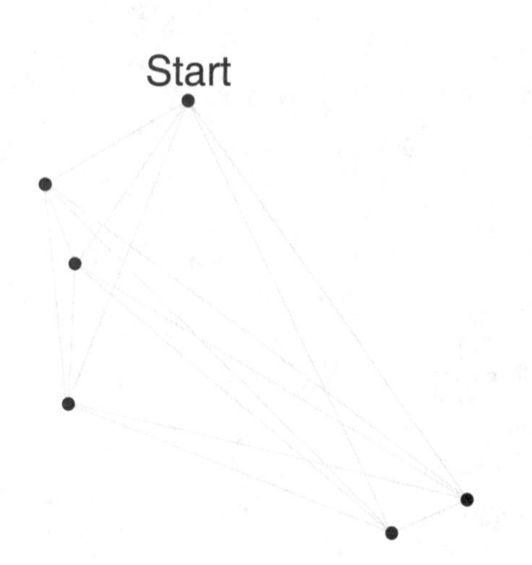

37. **Distances:**
3 cm, 5 cm, 3 cm, 1 cm, 5 cm.

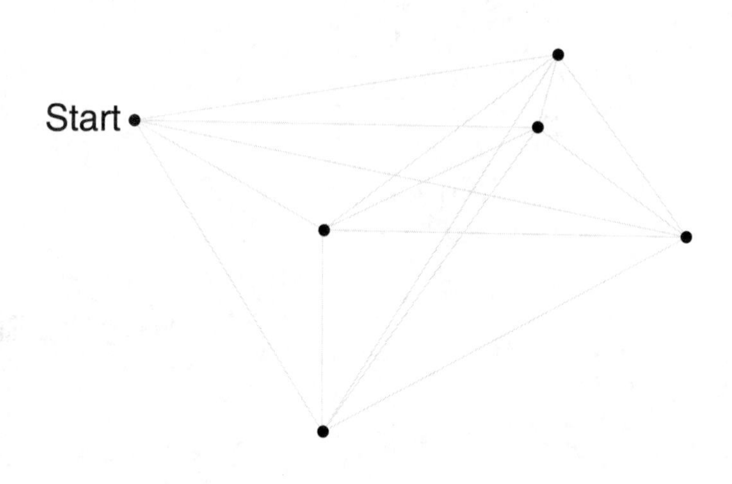

38. ★ **Distances:**
2 cm, 3 cm, 4 cm, 3 cm, 4 cm.

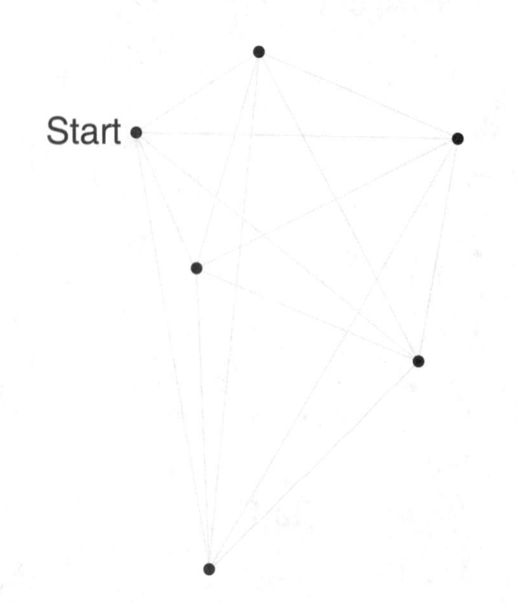

39. ★ **Distances:**
3 cm, 3 cm, 1 cm, 3 cm, 5 cm.

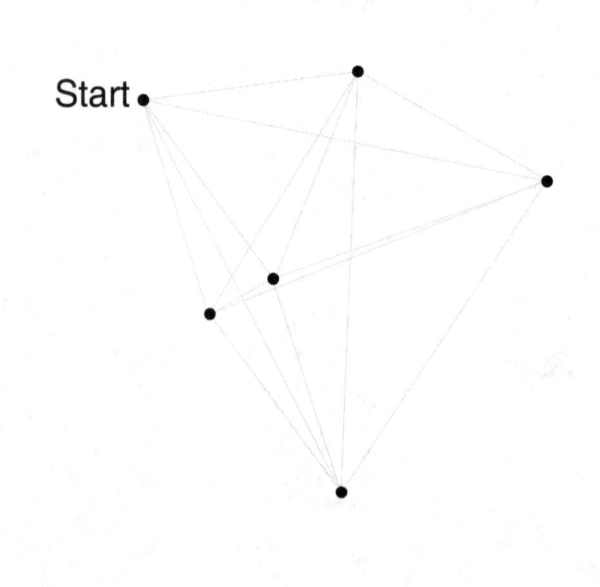

PRACTICE Solve each Measure-Maze below.
In these mazes, you are not given the Start.

40. ★ **Distances:**
5 cm, 2 cm, 1 cm, 5 cm.

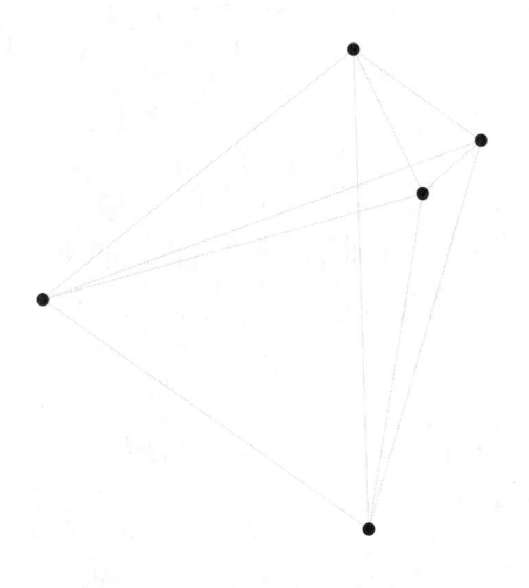

41. ★ **Distances:**
4 cm, 7 cm, 7 cm, 2 cm.

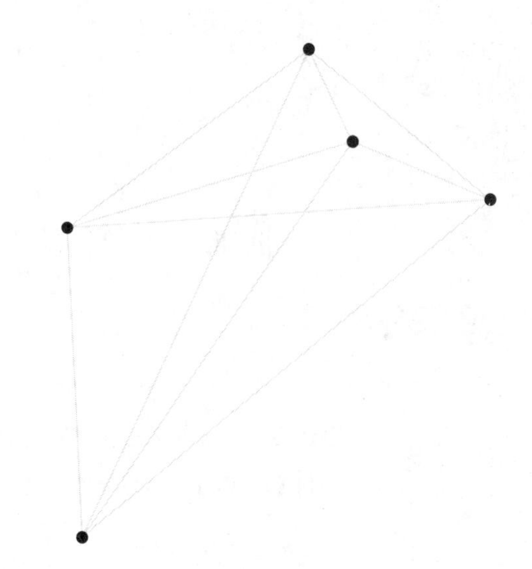

42. ★ **Distances:**
5 cm, 5 cm, 4 cm, 5 cm, 3 cm.

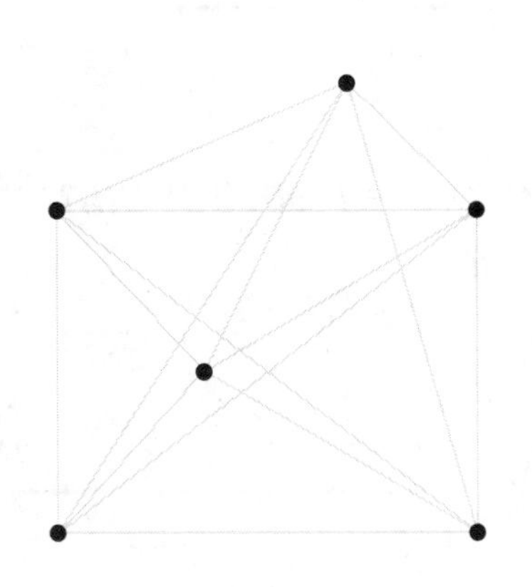

43. ★ **Distances:**
6 cm, 5 cm, 6 cm, 3 cm, 3 cm.

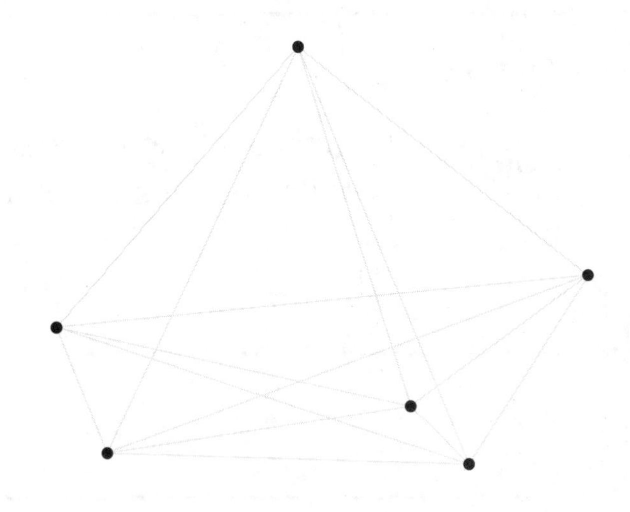

EXAMPLE To the nearest inch, how long is the Popsicle stick below?

The arrow on the ruler below points to the mark that is halfway between 4 and 5 inches. Since the Popsicle stick does not reach the arrow, it is closer to 4 inches than to 5 inches.

So, to the nearest inch, the Popsicle stick is about **4 inches** long.

PRACTICE Label the length of each line below to the ***nearest inch***. Write your answer on top of each line.

44. ______________________________

45. ____________________

46. ____

PRACTICE Label the length of each line below to the ***nearest centimeter***. Write your answer on top of each line.

47. ______________________________

48. ____________________

49. ____

PRACTICE | Answer each question below.

50. Draw a line below that is 15 centimeters long. Then, find the length of the line you drew to the ***nearest inch***.

50. ______ in

51. Draw a line below that is 5 inches long. Then, find the length of the line you drew to the ***nearest centimeter***.

51. ______ cm

52. Write the following lengths in order from shortest to longest:

9 cm, 12 cm, 4 in, 5 in.

52. ______, ______, ______, ______

53. Alex draws lines that are 10, 11, 12, 13, 14, and 15 centimeters long. ***How many*** of Alex's lines are between 4 and 5 inches long?

53. ______

54. ★ 2 inches is very close to 5 centimeters, so 4 inches is about 10 centimeters. Circle the length below that is closest to 40 inches.

16 centimeters   50 centimeters   80 centimeters   100 centimeters

Inches and centimeters are used to measure short lengths and distances. For longer measurements, we usually use longer units.

In the United States, the most common units of length are inches, feet, yards, and miles. These are called ***customary units***.

There are 12 inches in 1 ***foot*** (ft).
This page is a little less than 1 foot from top to bottom.

There are 3 feet in 1 ***yard*** (yd).
The front door of a house is usually about 1 yard wide.

There are 1,760 yards (which is 5,280 feet) in 1 ***mile*** (mi).
It takes 20-30 minutes to walk a mile, but a car can travel a mile in about 1 minute on the highway.

PRACTICE | Draw a line to connect each item on the left to the measurement on the right that is the best match.

55. The height of a refrigerator. • 9 inches

56. The height of the ceiling in a home. • 15 miles

57. The length of a community pool. • 25 yards

58. The width of a square baking dish. • 9 feet

59. The distance between nearby towns. • 6 feet

Outside of the United States, common units of length are centimeters, meters, and kilometers.

These are called ***metric units***.

There are 100 centimeters in 1 ***meter*** (m).
A meter is a little longer than a yard, about the height of a high kitchen counter or a typical 4-year-old child.

There are 1,000 meters in 1 ***kilometer*** (km).
A kilometer is shorter than a mile. It takes 12-18 minutes to walk a kilometer.

PRACTICE | Draw a line to connect each item on the left to the measurement on the right that is the best match.

60. The length of a clothes hanger. • 15 centimeters

61. The length of a bed. • 2 kilometers

62. The length of a pen. • 2 meters

63. The length of an airport runway. • 8 meters

64. The height of a flagpole. • 45 centimeters

EXAMPLE | How many inches equal 3 feet?

Since 1 foot equals 12 inches,
3 feet equals $12+12+12=\mathbf{36}$ inches.

EXAMPLE | How many yards equal 15 feet?

Since 1 yard is 3 feet, we count how many times we add 3 feet to get 15 feet.

Adding five 3's gives us
$3+3+3+3+3=15$. So, **5** yards equals
$3+3+3+3+3=15$ feet.

PRACTICE | Answer each question below.

65. How many feet long is a rope that is 4 yards long?

65 ______ ft

66. Marylou's desk is 60 inches wide.
How many feet wide is her desk?

66. ______ ft

67. Toby Termite walks once around a square that has sides that are 6 feet long. How many ***yards*** does Toby walk?

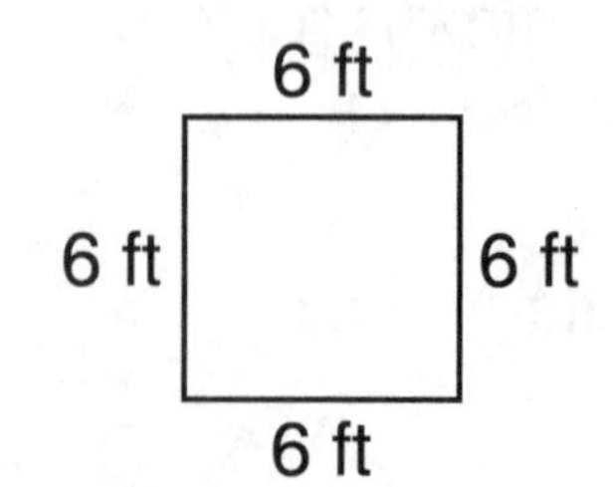

67. ______ yd

PRACTICE | Answer each question below.

68. How many centimeters are in 3 meters? **68.** ______ cm

69. How many inches are in 2 yards? **69.** ______ in

70. How many meters tall is a sign that is 600 centimeters tall? **70.** ______ m

71. Dar's boat is 20 feet long. Tog's boat is 6 yards long. How many feet longer is Dar's boat than Tog's boat? **71.** ______ ft

72. Gerb's tail is 4 feet long. Herb's tail is 55 inches long. How many inches longer is Herb's tail than Gerb's? **72.** ______ in

73. Tiki cuts a 1-meter string into five equal pieces. How many centimeters long is each piece? **73.** ______ cm

EXAMPLE | Greg is 4 feet 5 inches tall. What is Greg's height in inches?

The height "4 feet 5 inches" means that Greg is 5 inches more than 4 feet tall.

4 feet is $12+12+12+12=48$ inches.

So, 4 feet 5 inches is $48+5=$ **53** inches.

PRACTICE | Fill each circle below with <, >, or =.

74. 2 ft 3 in ◯ 30 in

75. 40 in ◯ 3 ft 6 in

76. 15 in ◯ 1 ft 3 in

77. 10 ft 1 in ◯ 101 in

78. 7 ft 8 in ◯ 8 ft 7 in

79. 2 yd ◯ 5 ft 11 in

80. How many inches longer than 3 feet 3 inches is 4 feet 4 inches?

80. ______ in

81. How many inches longer than 3 feet 6 inches is 6 feet 3 inches?

81. ______ in

EXAMPLE | Write 43 inches using feet and inches.

3 feet is 12+12+12=36 inches.
4 feet is 12+12+12+12=48 inches.
So, 43 inches is more than 3 feet but less than 4 feet.

Since 43−36=7, we know 43 inches is 7 inches more than 36 inches.

So, 43 inches is **3 feet 7 inches**.

When we use feet and inches to give a measurement...

...the number of inches should always be ***less than*** 12.

PRACTICE | Answer each question below in feet and inches.

82. Write 31 inches using feet and inches. **82.** _____ ft _____ in

83. Write 52 inches using feet and inches. **83.** _____ ft _____ in

84. Julian is 3 inches less than 5 feet tall.
Navin is 5 inches taller than Julian.
Write Navin's height using feet and inches. **84.** _____ ft _____ in

85. Brobdig writes that he is 14 feet 14 inches tall.
What is the correct way to write Brobdig's height using feet and inches? **85.** _____ ft _____ in

In a **Length Link** puzzle, we draw paths that connect squares with equal lengths as shown in the solved example below.

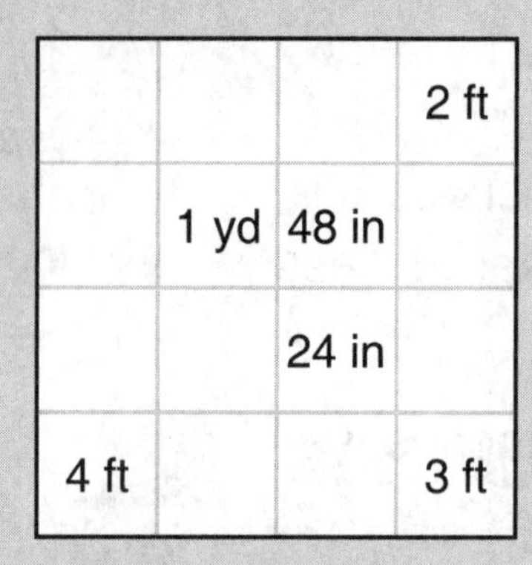

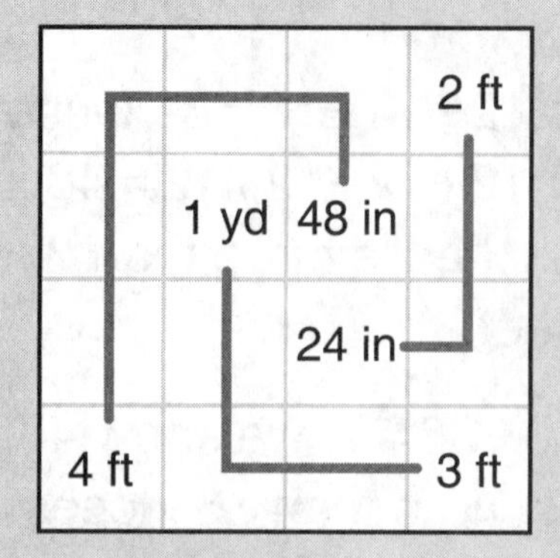

- Paths may not go diagonally.
- Paths begin and end at a length, but may not pass through a square that contains a length.
- Only one path may pass through each square.

PRACTICE | Solve each Length Link puzzle below.

86.

5 ft			
		6 ft	
		12 ft	
2 yd	4 yd	60 in	

87.

			1 ft
	1 yd		
			12 in
5 yd	36 in		15 ft

88.

			12 in
	24 in	3 ft	
		2 ft	
1 yd	1 ft		

89.

9 ft			
		48 in	
3 yd			12 ft
4 yd			4 ft

Print more Length Link puzzles at BeastAcademy.com.

PRACTICE | Solve each Length Link puzzle below.

90.

	36 in		3 ft	
	6 ft			18 ft
		6 yd		2 yd

91.

				15 ft
		24 in		
9 ft		3 yd		
2 ft		5 yd		

92.

	3 ft	48 in		1 yd
		60 in		
		4 ft		
			5 ft	

93.

12 in				
		36 in		
4 yd				
1 yd		12 ft		1 ft

94.

21 ft				120 in
	72 in			2 yd
	7 yd	10 ft		

95.

				12 ft
		30 ft		
			4 yd	
36 in	10 yd			1 yd

PRACTICE | Solve each Length Link puzzle below.

96.

	6 ft		2 yd	
5 yd	36 in			
		4 ft	3 ft	
48 in	15 ft			

97.

9 ft			18 ft	3 yd
24 in				
				60 in
		6 yd		
2 ft				5 ft

98. ★

6 ft			24 ft	10 yd	
8 yd		48 in			
				4 ft	
	360 in				
					2 yd

99. ★

					36 in
			240 in		
		9 yd	72 in		
				3 ft	
		20 ft		27 ft	2 yd

EXAMPLE Ian stacks a bookcase that is 3 ft 11 in tall on top of a desk that is 2 ft 8 in tall. What is the distance from the top of the bookcase to the floor in feet and inches?

To find the distance, we add the heights: 3 ft 11 in + 2 ft 8 in. We can add the feet and inches separately.

3 ft + 2 ft = 5 ft.   11 in + 8 in = 19 in.

So, 3 ft 11 in + 2 ft 8 in = 5 ft + 19 in.

19 inches is 1 ft + 7 in. So, <u>5 ft + 19 in</u> equals <u>5 ft + 1 ft + 7 in</u>, which is **6 ft 7 in**.

We often stack the units when adding lengths as shown below.

```
   3 ft 11 in
+  2 ft  8 in
-------------
   5ft  19in   (crossed out)
   6ft   7in
```

Remember, the number of inches in a mixed measure should always be ***less than*** 12.

PRACTICE Add each pair of lengths below. Circle your final answer.

100.
```
   5 ft  7 in
+  2 ft  3 in
-------------
```

101.
```
   3 ft  8 in
+  6 ft  5 in
-------------
```

102.
```
   4 ft  7 in
+  1 ft  9 in
-------------
```

103. ★ Barney Beetle walks once around a triangle that has sides that are 3 ft 9 in long. How many feet and inches does Barney walk?

103. _____ ft _____ in

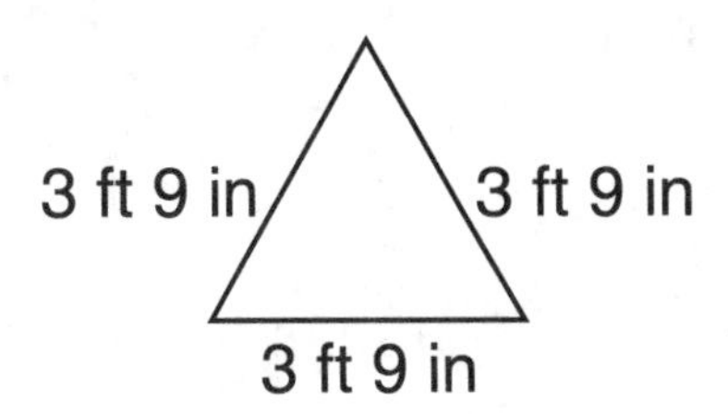

EXAMPLE Maris cuts 1 ft 7 in off the top of a post that is 4 ft 3 in tall. How tall is the remaining post in feet and inches?

To find the new height, we subtract 4 ft 3 in − 1 ft 7 in.

We can subtract the feet and inches separately. Since we can't take 7 inches from 3 inches, we rewrite 4 ft 3 in by breaking 1 foot into 12 inches.

4 ft 3 in is the same as 3 ft 15 in.

3 ft 15 in
~~4 ft~~ ~~3 in~~
− 1 ft 7 in

Now, 3 ft − 1 ft = 2 ft, and 15 in − 7 in = 8 in.

So, 4 ft 3 in − 1 ft 7 in = **2 ft 8 in**.

3 ft 15 in
~~4 ft~~ ~~3 in~~
− 1 ft 7 in
2 ft 8 in

PRACTICE Subtract each pair of lengths below. Circle your final answer.

104. 3 ft 7 in − 1 ft 3 in

105. 6 ft 1 in − 4 ft 7 in

106. 8 ft 7 in − 2 ft 11 in

107. Big Red is 5 ft 2 in tall. Nellie is 2 ft 5 in tall. How much taller is Big Red than Nellie?

107. _____ ft _____ in

PRACTICE | Answer each question below.

108. How tall is a stack of 7 concrete blocks if each block is 8 inches tall?

108. _____ ft _____ in

109. Ernie Inchworm's office is 7 feet from his home. Ernie has traveled 4 ft 5 in from home to his office. How much farther does he have to go?

109. _____ ft _____ in

110. Marie cuts 20 inches from a rope that is 20 feet long. What is the length of the remaining rope in feet and inches?

110. _____ ft _____ in

111. Compute 5 ft 8 in + 4 ft 7 in − 6 ft 5 in.

111. _____ ft _____ in

+

PRACTICE | Answer each question below.

112. ★ How many inches longer is 11 feet than 11 inches?

112. ______ in

113. ★ Subtract 5 feet from 4 yards. Give your answer in inches.

113. ______ in

114. ★ A stop sign has 8 sides, all the same length. The distance around the outside of the stop sign shown is 2 yards. What is the length in inches of one side of the stop sign?

114. ______ in

115. ★ There are four tiny plus signs on this page: three below, and one in the top-left part of the page. It takes six lines to connect every plus sign to every other plus sign. What is the total length in centimeters of these six lines?

115. ______ cm

+

+

+

PRACTICE | Answer each question below.

116. How many centimeters longer is the long line than the short line below?

116. _____ cm

117. ★ Kirby Cricket travels 35 inches with every hop he takes. How many ***feet*** will Kirby travel with 12 hops?

117. _____ ft

118. ★ Each day last week, it snowed one more inch than it did the day before. On Sunday it snowed 9 inches, on Monday it snowed 10 inches, on Tuesday it snowed 11 inches, and so on. On what day did the ***total*** snowfall for the week pass 5 feet?

118. ___________

119. ★ Four of the same small rectangles are put together to make a big rectangle as shown. The long side of each ***small*** rectangle is 1 foot. How many inches long is the long side of the ***big*** rectangle?

1 ft

119. _____ in

CHAPTER 8
Strategies

Use this Practice book with Guide 2C from BeastAcademy.com.

Recommended Sequence:

Book	**Pages:**
Guide:	44-59
Practice:	37-56
Guide:	60-69
Practice:	57-69

You may also read the entire chapter in the Guide before beginning the Practice chapter.

EXAMPLE | Evaluate 13−4+19−7.

$$13-4+19-7$$
$$= 9+19-7$$
$$= 28-7$$
$$= \mathbf{21}.$$

PRACTICE | Evaluate each expression below.

1. 9+8−3 = ________

2. 15−6+4 = ________

3. 11−8+4−5 = ________

4. 8−1+2+6 = ________

5. 47+23−14 = ________

6. 100−68+32 = ________

7. 87−37+25−11 = ________

8. 24+36+18−31 = ________

9. 18+12−7−8+2 = ________

10. 350−40−200−40+110 = ________

EXAMPLE | Add $59+84+41$.

We can start by adding $59+84$. But it's easier to start by adding $59+41=100$. Then, $100+84=$ **184**.

$$\begin{aligned} & 59+84+41 \\ = & \ 100+84 \\ = & \ 184. \end{aligned}$$

PRACTICE | Find each sum below.

11. $32+49+51=$ ________

12. $17+44+23=$ ________

13. $25+39+25=$ ________

14. $13+18+17+12=$ ________

15. $111+89+16+24=$ ________

16. $35+35+65+65=$ ________

17. $126+32+74+118=$ ________

18. $12+245+99+118+105=$ ________

19. ★ $5+15+25+35+45+55+65+75+85+95=$ ________

EXAMPLE | Subtract $156-44-56$.

Subtracting 44 then 56 is the same as subtracting 56 then 44. So, $156-44-56$ is equal to $156-56-44$.

Since it is easier to start by subtracting 56, we compute $156-56-44$, which gives us $100-44=\mathbf{56}$.

– or –

Subtracting 44 then 56, we subtract a total of 100. So, $156-44-56$ is equal to $156-100=\mathbf{56}$.

Be careful when changing the order of subtraction.

We can't reorder the number we subtract from. $156-100$ is ***not*** equal to $100-156$.

PRACTICE | Solve each problem below.

20. $86-9-56=$ ________

21. $579-25-25=$ ________

22. $276-39-26=$ ________

23. $353-37-63=$ ________

24. Circle every expression below that is equal to $88-12-28$.

$88-28+12$ $88+12-28$ $88-28-12$ $88+40$ $88-40$

25. Circle every expression below that is equal to $155-99$.

$155-100+1$ $155-55-44$ $155-55+44$ $156-100$

EXAMPLE Morgan has $186. She earns $145, then spends $86. How many dollars does Morgan have now?

Morgan has 186+145−86 dollars:

$$\begin{aligned} & 186+145-86 \\ =\ & 331-86 \\ =\ & 245. \end{aligned}$$

So, Morgan has **245** dollars.

— or —

Morgan starts with $186. Earning $145 then spending $86 leaves her with the same amount of money as spending $86 then earning $145.

Since subtracting 86 first is easier, we compute 186−86+145=100+145=245.

So, Morgan has **245** dollars.

PRACTICE Solve each word problem below.

26. A quilt has 144 squares. Elmer adds 180 squares to the quilt, then removes 44 of them. How many squares does the quilt have now?

26. ________

27. A pack of gorillamas has a pile of 125 bananas. They eat 48 of the bananas. Then, they add 25 bananas to the pile. How many bananas are in the pile now?

27. ________

28. Buster has a flock of 141 flamingoats. He sells 23 of his flamingoats. Then, 31 flamingoats wander off, but 20 return. How many flamingoats are in Buster's flock now?

28. ________

EXAMPLE | Evaluate $213+89-12$.

Adding 89 then subtracting 12 gives the same result as subtracting 12 then adding 89.

So, $213+89-12$ is equal to $213-12+89$.

We have $213-12+89=201+89=\mathbf{290}$.

When you rearrange addition and subtraction...

...***always*** keep the + and − signs with the numbers you are adding and subtracting.

PRACTICE | Solve each problem below.

29. Circle the expression below that is equal to $32+39-32$.

$32+32-39$ $\quad$ $32-32+39$ $\quad$ $32-32-39$ $\quad$ $32+32+39$

30. Circle the expression below that is equal to $55-16+45$.

$55+45+16$ $\quad$ $55-45-16$ $\quad$ $55+45-16$ $\quad$ $55-45+16$

31. Circle the expression below that is equal to $44-18+44$.

$44+44-18$ $\quad$ $44-44-18$ $\quad$ $44+44+18$ $\quad$ $44-44+18$

32. Evaluate each expression below.

$32+39-32=$ _____ $\quad$ $55-16+45=$ _____ $\quad$ $44-18+44=$ _____

PRACTICE | Solve each problem below.

33. Circle the ***two*** expressions below that are equal to $127-73+33-27$.

$127+33-73-27$ $\quad$ $127-27+73-33$

$127-33+73-27$ $\quad$ $127-27+33-73$

34. Circle the ***two*** expressions below that are equal to $136+20+64-36$.

$136+36-64+20$ $\quad$ $136-36+64+20$

$136+64-20-36$ $\quad$ $136+64-36+20$

35. Fill each ◯ below with + or − to make the equations true.

$111-36+55=111$ ◯ 55 ◯ 36

$168-45+132=168$ ◯ 132 ◯ 45

$158+17-58+42=158$ ◯ 58 ◯ 42 ◯ 17

36. Fill each ☐ below with a number to make the equations true.

$78+49-68=78-$ ☐ $+$ ☐

$135-44-35+22=135-$ ☐ $-44+$ ☐

$91-38-12+9=91+$ ☐ $-$ ☐ -38

PRACTICE | Evaluate each expression below.

37. $222+44-222=$ ________

38. $222-44+222=$ ________

39. $235+68-35=$ ________

40. $235-68+35=$ ________

41. $95+14-30+5=$ ________

42. $88+56-44-25=$ ________

43. $142-34+123-42=$ ________

44. $60+65-30+35=$ ________

45. $235+88-135-57=$ ________

46. $81-37-43+19=$ ________

In a **Crosstile** puzzle, fill the blank squares with the given entries to make true equations reading from left-to-right and top-to-bottom.

EXAMPLE | Use −1, −2, +3, and +4 to complete the Crosstile puzzle on the right.

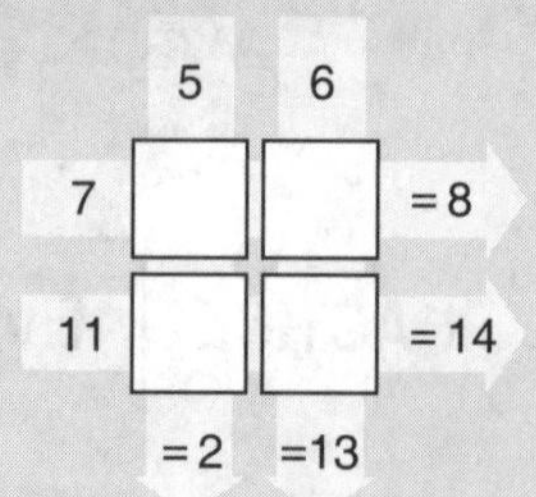

In the left column, to get from 5 to 2, we subtract 3. We can only subtract 3 using −2 and −1.

In the bottom row, to get from 11 to 14, we add 3. We can only add 3 using +4 and −1.

−1 & −2
5 6
7 =8
−1 & +4 11 −1 =14
=2 =13

Since −1 must go in the left column and in the bottom row, we place −1 in the bottom-left square.

We complete the puzzle by placing −2, +4, and +3 as shown.

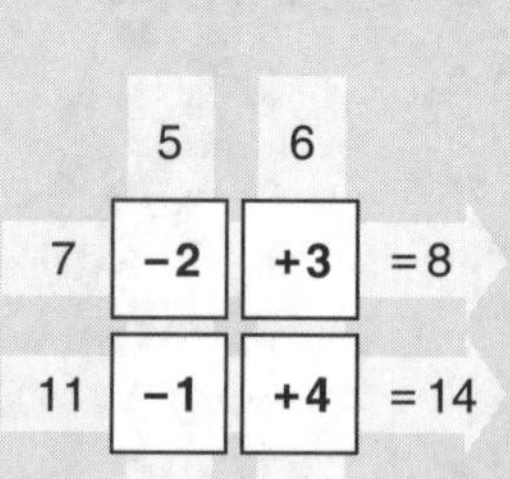

We then check our work:

$7-2+3=8$. ✓ $5-2-1=2$. ✓
$11-1+4=14$. ✓ $6+3+4=13$. ✓

PRACTICE | Solve each Crosstile puzzle below.

47. **Entries:** +1, +2, +3, +5

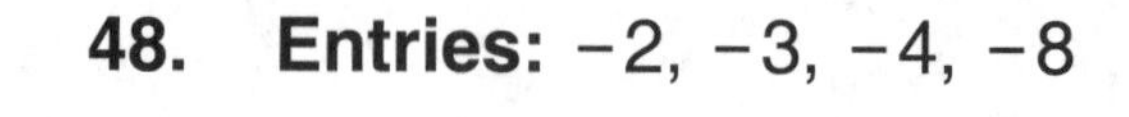

48. **Entries:** −2, −3, −4, −8

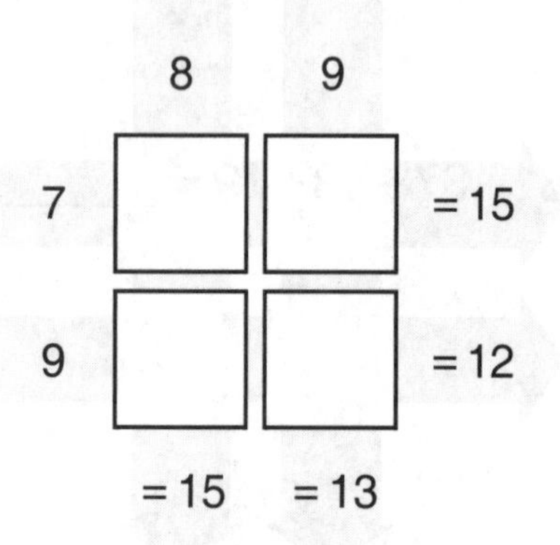

21 24
15 =5
10 =3
=10 =18

PRACTICE | Solve each Crosstile puzzle below.

49. **Entries:** −1, +1, −5, +5

	10	10	
10			= 16
10			= 4
	= 6	= 14	

50. **Entries:** +1, +3, −5, +5

	18	3	
7			= 11
9			= 9
	= 14	= 11	

51. **Entries:** +2, −4, −4, +8

	10	10	
10			= 8
10			= 14
	= 14	= 8	

52. **Entries:** +1, −2, −3, −6

	22	25	
24			= 22
28			= 20
	= 21	= 16	

53. **Entries:** +3, −4, +5, −6

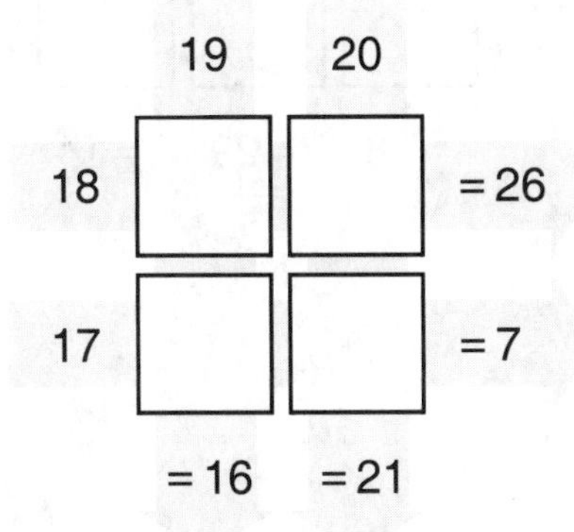

54. **Entries:** +11, −22, −33, +44

	55	55	
55			= 44
55			= 66
	= 77	= 33	

PRACTICE | Solve each Crosstile puzzle below.

55. **Entries:** −1, +1, −2, +2, −3, +3

	10	10	10	
10				= 16
10				= 4
	= 12	= 10	= 8	

56. **Entries:** −1, −1, −1, +1, +1, +1

	4	3	2	
5				= 4
6				= 7
	= 4	= 1	= 4	

57. ★ **Entries:** −1, +2, −3, +4, −5, +6

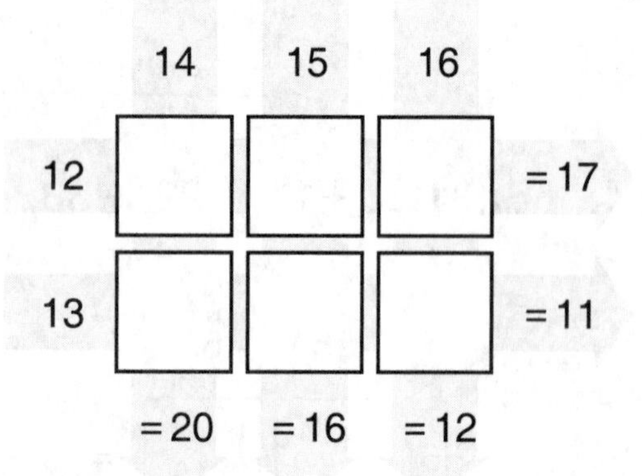

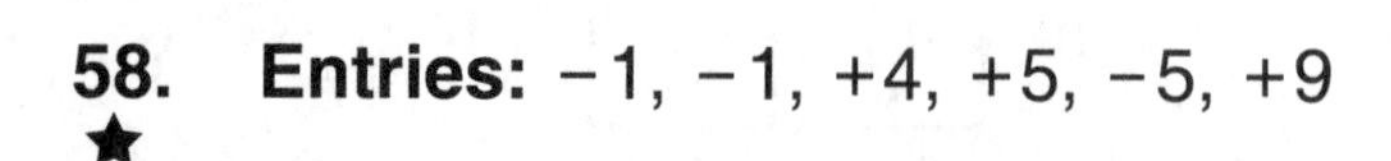

58. ★ **Entries:** −1, −1, +4, +5, −5, +9

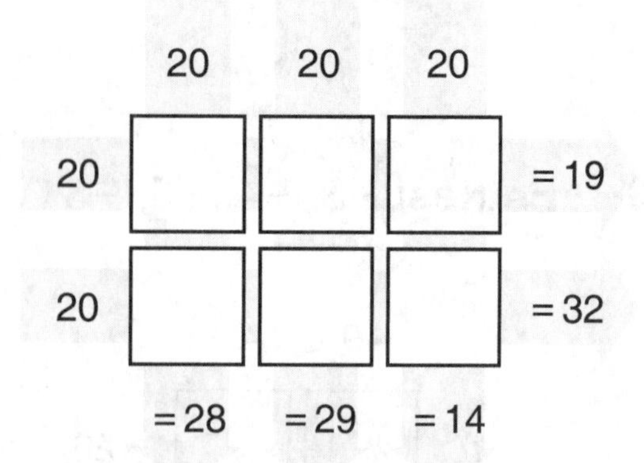

PRACTICE | Solve each Crosstile puzzle below.

59. ★ **Entries:** +1, +3, −3, +4, −5, +8

	12	20	8	
15				= 16
18				= 25
	= 13	= 29	= 6	

60. ★ **Entries:** −1, −1, +1, +1, −2, +2

	4	4	4	
4				= 2
4				= 6
	= 1	= 4	= 7	

61. ★★ **Entries:** +1, +2, −3, −5, −6, −8

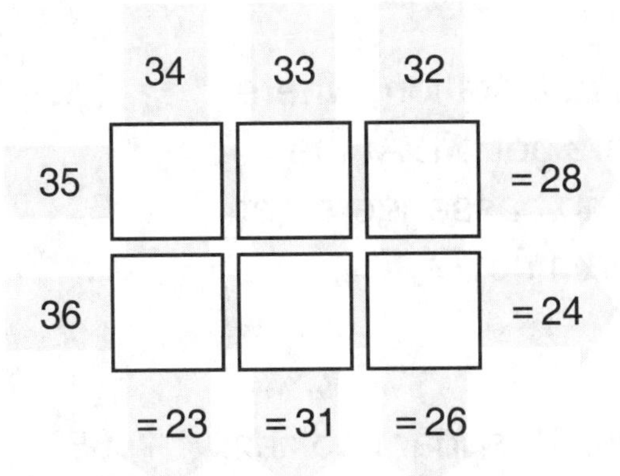

62. ★★ **Entries:** +0, +1, −2, +2, −3, +3

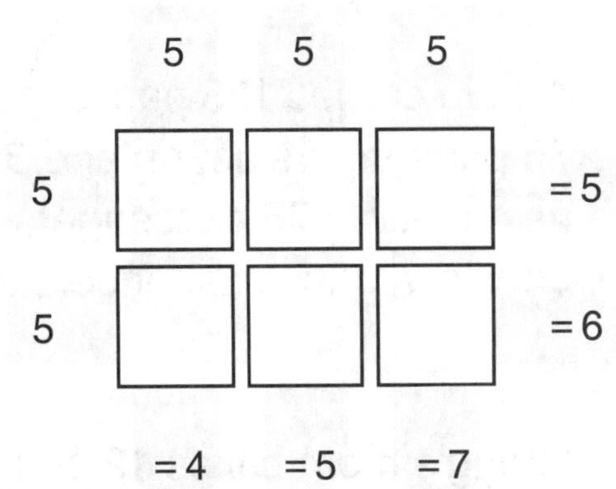

STRATEGIES

Canceling

EXAMPLE Captain Kraken has 555 gold coins. He finds a treasure of 192 coins, then spends 192 coins to pay for ship repairs. How many gold coins does Captain Kraken have now?

Captain Kraken now has 555+192−192 coins:

$$\begin{aligned} &555+192-192 \\ =\ &747-192 \\ =\ &555. \end{aligned}$$

So, Captain Kraken now has **555** coins.

– or –

Since Captain Kraken spent the same number of coins that he found, he ends up with the same number of coins that he started with.

So, Captain Kraken now has **555** coins.

PRACTICE Solve each word problem below.

63. An elevator starts on floor 15. It goes up 78 floors, then down 78 floors. What floor is the elevator on now?

63. ________

64. A train carries 136 passengers. It stops at a station, where 75 passengers get on and 36 passengers get off. At the next station, 75 passengers get off and no passengers get on. How many passengers are on the train now?

64. ________

65. Grogg's mom bakes 12 oatmeal cookies, 15 sugar cookies, and 24 chocolate chip cookies. Grogg eats 6 oatmeal cookies, 12 sugar cookies, and 15 chocolate chip cookies. How many cookies are left?

65. ________

EXAMPLE | Evaluate $54+37+26-37$.

We can switch $+26$ and -37.

Then, adding 37 and subtracting 37 is the same as doing nothing. We say that $+37$ and -37 ***cancel*** each other.

$$54+37+26-37 = 54+37-37+26$$
$$= 54+26$$
$$= \mathbf{80}.$$

We can cancel $+37$ and -37 without rearranging as shown below.

$$54+37+26-37 = 54+26$$
$$= \mathbf{80}.$$

PRACTICE | Fill in the blanks to solve each problem below.

66. $53+43+58-58$

$= 53+$ ☐

$=$ ☐

67. $19-11+11+7$

$= 19+$ ☐

$=$ ☐

68. $36+43-32-43$

$= 36-$ ☐

$=$ ☐

69. $13+88+35-88 =$ ________

70. $114+112-113-112 =$ ________

71. $80+34-79-34+67 =$ ________

72. $99+88+33-88+33 =$ ________

73. $10+9+8-7-8-9+6+8+7+6 =$ ________

74. $23+32+22-33+23-22+33-32 =$ ________

EXAMPLE Ms. Q. has 48 colored pencils in her art bucket. She adds a new pack of 24 pencils to the bucket and removes 25 of the old pencils. How many pencils are left in the bucket?

The bucket now has $48+24-25$ pencils:

$$\begin{aligned} & 48+24-25 \\ = & \ 72-25 \\ = & \ 47. \end{aligned}$$

So, the bucket now has **47** pencils.

– or –

Ms. Q. added 24 pencils to the bucket, then removed 25. So, Ms. Q. removed 1 more pencil than she added. This means that the number of pencils in the bucket decreased by 1.

So, the bucket now has $48-1=$ **47** pencils.

PRACTICE Solve each word problem below.

75. There are 55 bags of potato chips on the display rack at a deli. The deli sells 36 bags of chips, then adds 38 more bags to the display rack. How many bags are now on the rack?

75. ________

76. Blorta has 120 flowers in her garden. Alligophers eat 78 of her flowers, but 75 of the flowers grow back. How many flowers does Blorta have now?

76. ________

77. In the game of Thonk, players earn 63 points for every clonk, but lose 62 points for every bonk. During a game, Winnie gets 3 clonks and 3 bonks. How many points does Winnie have?

77. ________

EXAMPLE | Evaluate $73+34-35$.

The $+34$ and the -35 ***almost*** cancel.

We subtract 1 more than we add. So, adding 34 then subtracting 35 is the same as subtracting 1.

$$\begin{aligned}73+34-35 &= 73-1\\ &= \mathbf{72}.\end{aligned}$$

PRACTICE | Fill in the blanks to solve each problem below.

78. $45+100-99$

$= 45+\square$

$= \square$

79. $72-55+60$

$= 72+\square$

$= \square$

80. $84-36+33$

$= 84-\square$

$= \square$

81. $68+55-50$

$= 68+\square$

$= \square$

82. $436-79+89$

$= 436+\square$

$= \square$

83. $839-287+387$

$= 839+\square$

$= \square$

84. $58+80-76 =$ ________

85. $49-33+99+33-100 =$ ________

86. ★ $100-11+12-13+14-15+16-17+18-19+20 =$ ________

STRATEGIES

All At Once

EXAMPLE | Evaluate 123−7+75−13+25.

Subtracting 7 then subtracting 13 is the same as subtracting 20. Adding 75 then adding 25 is the same as adding 100.

So, 123−7+75−13+25 is equal to 123−20+100.

$$\begin{aligned} & 123-7+75-13+25 \\ =\ & 123-20+100 \\ =\ & 103+100 \\ =\ & \mathbf{203}. \end{aligned}$$

PRACTICE | Fill in the blanks to solve each problem below.

87. 65−8+35−2

= 65+35−☐

= ☐

88. 53−24−26+38

= 53−☐+38

= ☐

89. 125+47−123+53

= 125−123+☐

= ☐

90. 50−12+20−18

= 50+20−☐

= ☐

91. 200−38+700−122

= 200+700−☐

= ☐

92. 19+15+15−6−6

= 19+☐−☐

= ☐

93. 134−19+33−21+17

= 134−☐+☐

= ☐

94. 7+81−11−33+9

= 7+☐−☐

= ☐

95. 80−19−11+22+23

= 80−☐+☐

= ☐

PRACTICE | Fill the boxes below to make each equation true.

96. $28 + 19 - \square = 28$

97. $789 + \square - 789 = 987$

98. $64 - 27 + \square = 65$

99. $113 - \square + 113 = 200$

100. $226 - 77 + \square = 224$

101. ★ $75 - 28 + 75 - \square = 100$

102. ★ $77 + 144 - 12 + \square - 145 = 77$

103. ★ $200 - 26 - \square + 27 + 84 = 203$

In an **Equation Path** puzzle, the goal is to trace a path through the grid to create a true equation. Start in the top-left corner and end with the bottom-right number outside the grid.

- The path can only go up, down, left, or right. (No diagonals.)
- The path may not cross the same square twice.

EXAMPLE | Solve the Equation Path puzzle on the right.

10	+3	−3	
+3	+2	+20	
−2	+30	−2	= 60

To get from 10 to 60, we need to add a total of $60-10=50$. To add 50, our path must cross +20 and +30. The other squares in our path must cancel.

The only way to do this is shown below.

10	+3	−3	
+3	+2	+20	
−2	+30	−2	= 60

We check our work:
$10+3-3+20+2+30-2=60$. ✓

PRACTICE | Solve each Equation Path puzzle below.

104.

15	+1	+1	
+1	+10	+1	
+10	+1	+10	= 28

105.

14	−1	−2	
+10	−2	−3	
+20	+30	−4	= 70

PRACTICE | Solve each Equation Path puzzle below.

106.

100	+17	−31
+3	−5	+3
+31	−17	−5

= 90

107.

26	+13	−25
−13	+25	+3
+3	+3	+3

= 32

108.

40	+2	−8
−40	+8	+10
+10	−2	+40

= 90

109.

6	+2	−1
+8	−8	+4
−2	+16	−4

= 0

110.

49	−10	+22
−10	−1	−1
+22	−10	−10

= 50

111.

19	+2	+2
+2	+4	−6
−6	−6	−6

= 17

PRACTICE | Solve each Equation Path puzzle below.

112. ★

13	+7	−1
−2	+7	−2
−3	+7	−3

= 26

113. ★

50	+6	+36
−7	+50	+8
+38	−37	−2

= 100

114. ★

100	+1	−30
−20	+2	+1
+2	−10	+50

= 114

115. ★

25	+1	+2
+7	−8	+3
+6	+5	+4

= 34

116. ★

26	+1	+20
+2	−5	+3
+20	+4	−20

= 28

117. ★

100	+11	−10
−20	+41	+61
−60	+21	−40

= 102

EXAMPLE | Evaluate $144-((57-17)-(12+18))$.

We start with the parentheses that are inside other parentheses.

Then, we evaluate the rest of the expression as shown.

$$\begin{aligned} & 144-((57-17)-(12+18)) \\ = \; & 144-(40-30) \\ = \; & 144-10 \\ = \; & \mathbf{134}. \end{aligned}$$

STRATEGIES

Parentheses Review

PRACTICE | Solve each problem below.

118. $128-(12+16)=$ ________

119. $55-(61-30)+26=$ ________

120. $50-(11+14-5)=$ ________

121. $99-(66-33)-(66-33)=$ ________

122. $27-(30-(9-5)-6)=$ ________

123. $99-((66-33)-(66-33))=$ ________

124. Place one pair of parentheses in the statement below to make it true.

$$22 - 7 - 6 + 5 - 4 = 0.$$

PRACTICE | Answer each question below.

125. Sam swam 17 laps in the morning and 17 laps in the afternoon. Myles swam 15 laps in the morning and 15 laps in the afternoon. Circle the ***two*** expressions below that show how many more laps Sam swam than Myles.

$(17-15)-(17-15)$ $\quad$ $(17+17)+(15+15)$

$(17+17)-(15+15)$ $\quad$ $(17-15)+(17-15)$

126. In the previous problem, how many more laps did Sam swim than Myles? **126.** ________

127. The Merkel brothers weigh 164 pounds and 138 pounds. The Samson sisters weigh 161 pounds and 133 pounds. Circle the ***two*** expressions below that show how many more pounds the Merkel brothers weigh than the Samson sisters.

$(164+138)-(161+133)$ $\quad$ $(164+161)-(138+133)$

$(164+138)+(161+133)$ $\quad$ $(164-161)+(138-133)$

128. In the previous problem, how many more pounds do the Merkel brothers weigh than the Samson sisters? **128.** ________

PRACTICE | Answer each question below.

129. Alex has 35 yellow candies, 23 red candies, and 14 green candies. Lizzie has 30 yellow candies, 21 red candies, and 12 green candies. Circle the ***two*** expressions below that show how many more candies Alex has than Lizzie.

$(35+23+14)+(30+21+12)$ $(35+23+14)-(30+21+12)$

$(35-30)+(23-21)+(14-12)$ $(35+30)-(23+21)-(14+12)$

130. In the previous problem, how many more candies does Alex have than Lizzie?

130. ________

131. Four mornings in a row, Grogg makes 15 waffles, eats 11 of them, and freezes the rest. Circle the ***two*** expressions below that show how many frozen waffles Grogg has after 4 days.

$(15+15+15+15)-(4+4+4+4)$ $(15+15+15+15)-(11+11+11+11)$

$(15-4)+(15-4)+(15-4)+(15-4)$ $(15-11)+(15-11)+(15-11)+(15-11)$

132. In the previous problem, how many frozen waffles does Grogg have after 4 days?

132. ________

EXAMPLE | Evaluate $50-(27+17)+(27+17)$.

Subtracting $(27+17)$ then adding $(27+17)$ is the same as doing nothing.

Since the expressions in the parentheses are the same, the subtraction and addition cancel.

$$50-(27+17)+(27+17)=\mathbf{50}.$$

Sometimes we can cancel expressions that are in parentheses.

PRACTICE | Evaluate each expression below.

133. $(38+57)-(38+57)=$ ________

134. $(19+23)-(23+19)=$ ________

135. $4+(9+5)-(9+5)=$ ________

136. $(7+6)-(6+7)+8=$ ________

137. $(8+3)+7+(8+3)-7=$ ________

138. $3-(11+7)+(7+11)+3=$ ________

139. $6+(7+8+9)+10-(9+8+7)=$ ________

140. ★ $(80+110)+50-(110+80)+(110-80)=$ ________

EXAMPLE | Evaluate $(48+47)-(46+45)$.

We can find the difference between $(48+47)$ and $(46+45)$ without computing either sum.

48 is 2 more than 46, and 47 is 2 more than 45. So, $48+47$ is $2+2=4$ more than $46+45$.

So, $(48+47)-(46+45)=\mathbf{4}$.

PRACTICE | Evaluate each expression below.

141. $(30+30)-(29+29)=$ ________

142. $(86+85)-(85+84)=$ ________

143. $(120+121)-(100+21)=$ ________

144. $(9+8+7)-(6+5+4)=$ ________

145. $(177+175+173)-(73+75+77)=$ ________

146. $(13+14+15+16+17)-(14+15+16)=$ ________

147. $(82+84+86+88+90)-(81+83+85+87+89)=$ ________

When we ***skip-count***, we add the same number over and over.

EXAMPLE | Skip-count by 10's to fill in the blanks below.

10, 20, _____, _____, _____, _____.

To skip-count by 10's, we add 10 over and over.

+10 +10 +10 +10 +10

10, 20, **30**, **40**, **50**, **60**.

PRACTICE | Fill the blanks in each skip-counting pattern below.

148. 2, 4, 6, _____, _____, _____, 14, _____, _____, _____.

149. 5, 10, _____, _____, _____, _____, _____, 40, _____, _____.

150. 20, 40, _____, _____, _____, _____, _____, _____, _____, _____.

151. 3, 6, 9, _____, _____, _____, _____, _____, _____, _____.

152. 4, 8, 12, _____, _____, _____, _____, _____, _____, _____.

153. _____, _____, _____, _____, 55, 66, 77, _____, _____, _____.

154. ★ 99, 198, _____, _____, _____, _____, _____, _____, _____, _____.

PRACTICE | Count the number of dots in each pattern below. Try to find ways to count the dots in groups by skip-counting.

155.

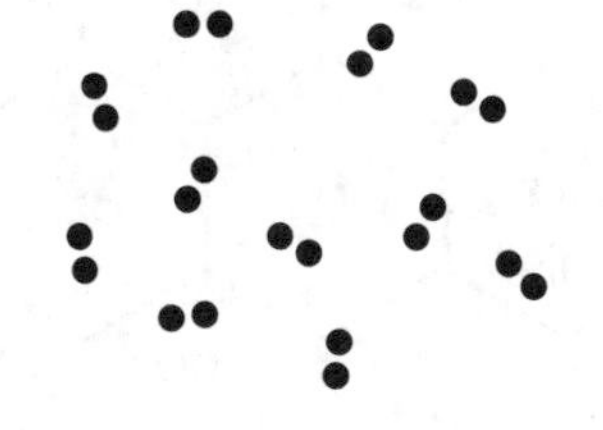

_____ dots

156.

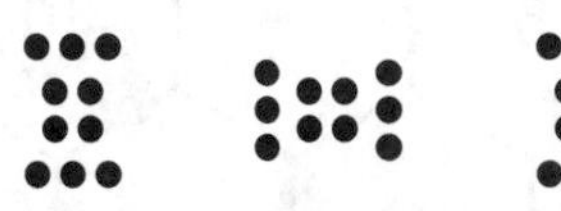

_____ dots

157.

_____ dots

158.

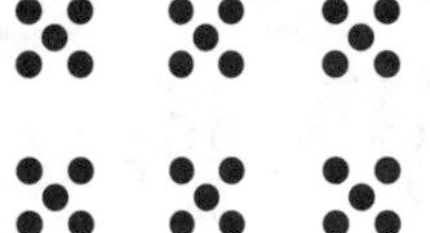

_____ dots

159.

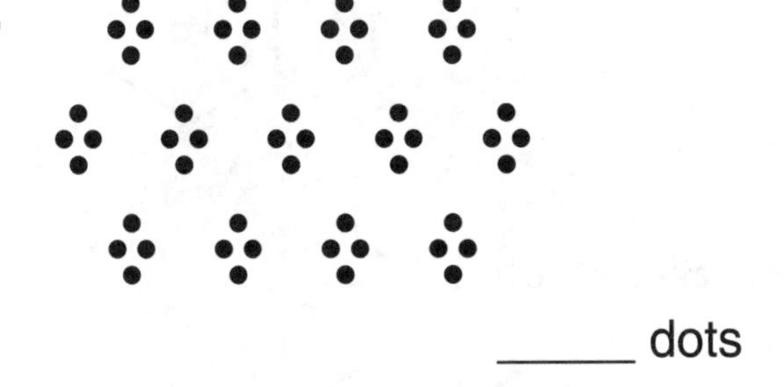

_____ dots

160.

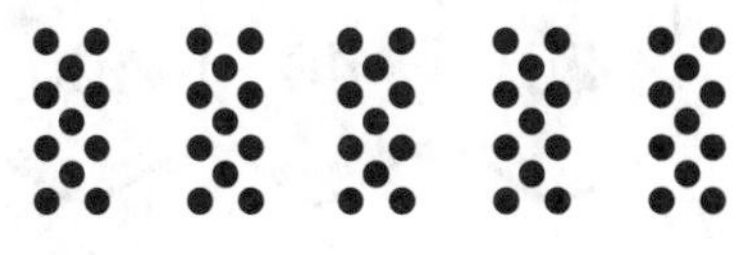

_____ dots

161.

_____ dots

162.

_____ dots

In a **Skip-Counting Honeycomb Path** puzzle, the goal is to fill every empty hexagon (⬡) with a number so that a skip-counting pattern forms a path that crosses every hexagon.

We skip-count by 5's to complete the Honeycomb Path below.

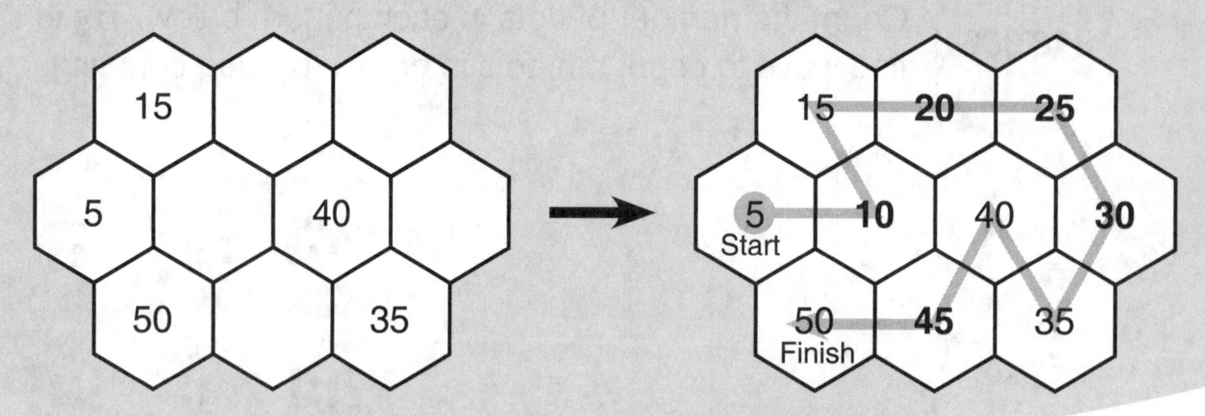

PRACTICE | Skip-count by the given number to solve each Honeycomb Path puzzle below.

163. Skip-count by 10's.

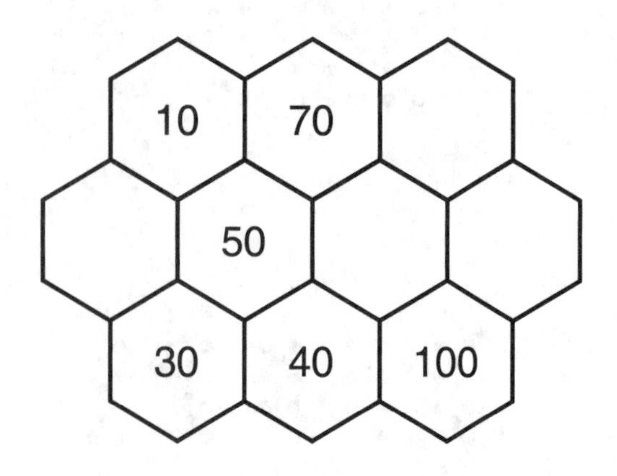

164. Skip-count by 2's.

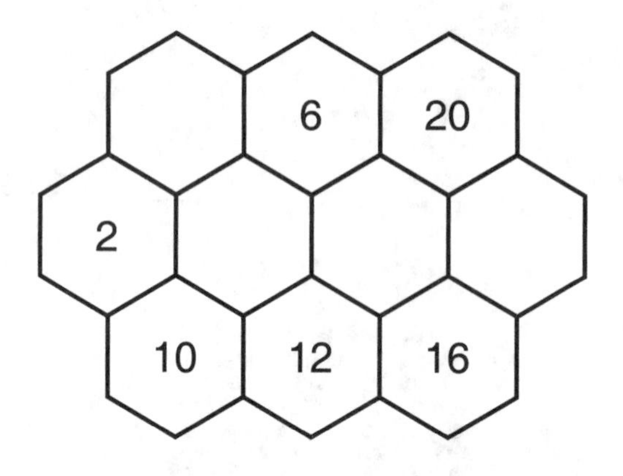

165. Skip-count by 3's.

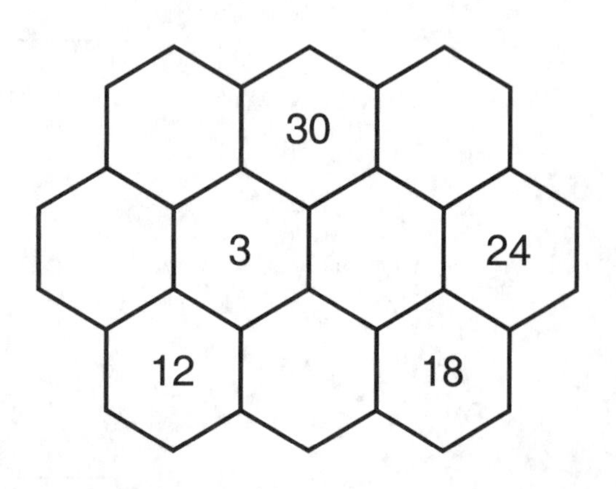

166. Skip-count by 5's.

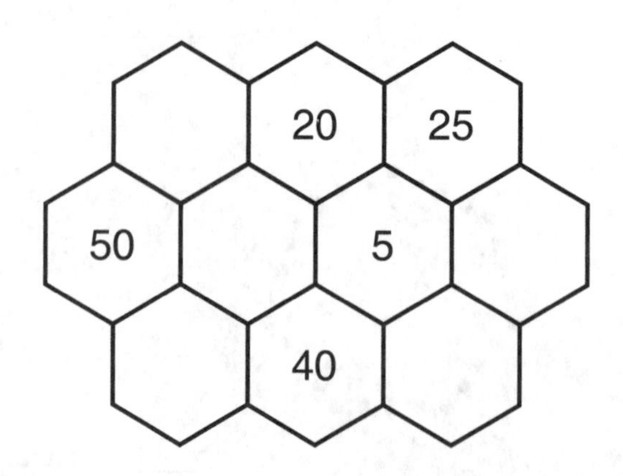

PRACTICE | Skip-count by the given number to solve each Honeycomb Path puzzle below.

167. Skip-count by 4's.

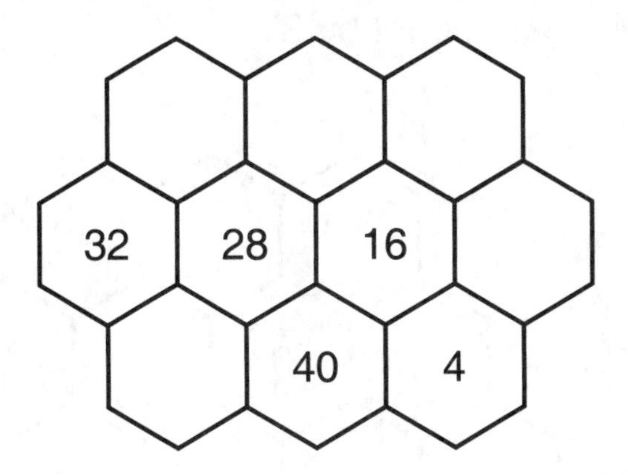

168. Skip-count by 20's.

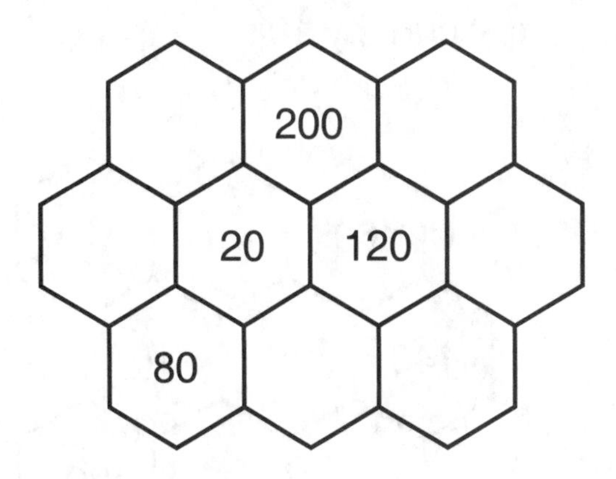

169. Skip-count by 2's.

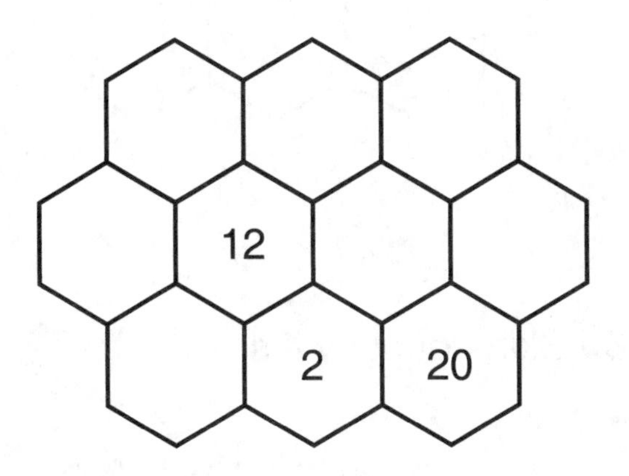

170. Skip-count by 50's.

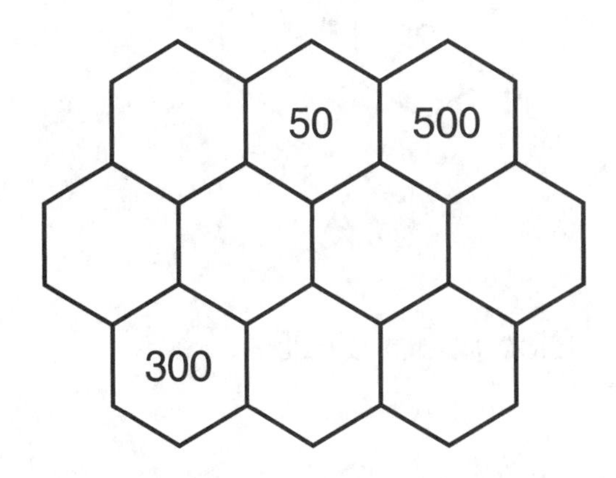

171. Skip-count by 11's.

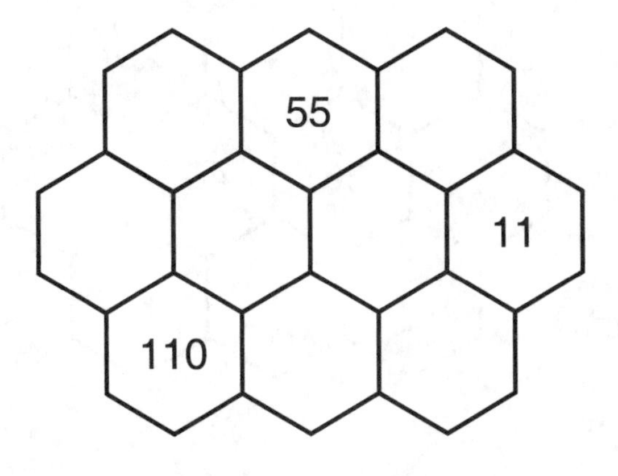

172. Skip-count by 9's.

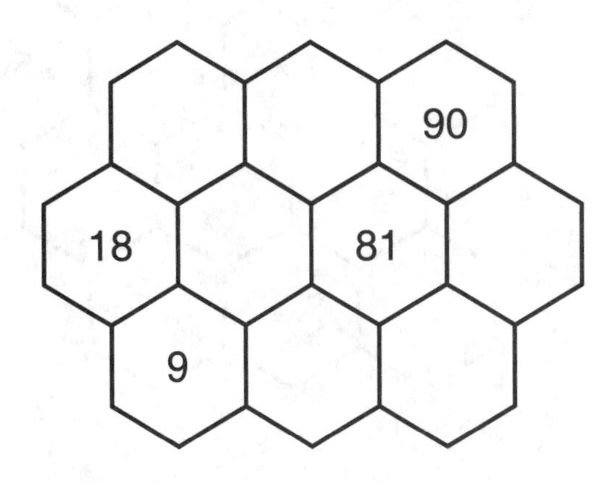

PRACTICE | Skip-count by the given number to solve each Honeycomb Path puzzle below.

173. Skip-count by 10's.

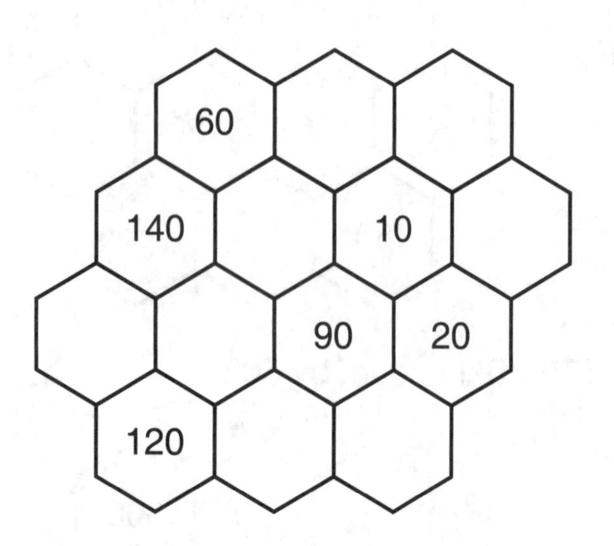

174. Skip-count by 3's.

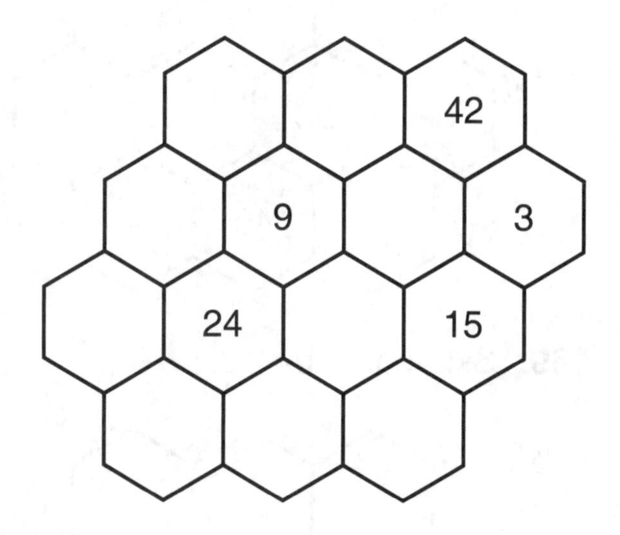

175. ★ Skip-count by 25's.

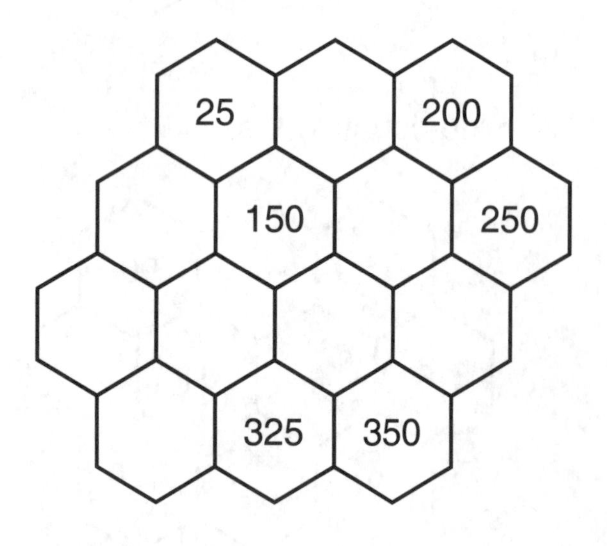

176. ★ Skip-count by 4's.

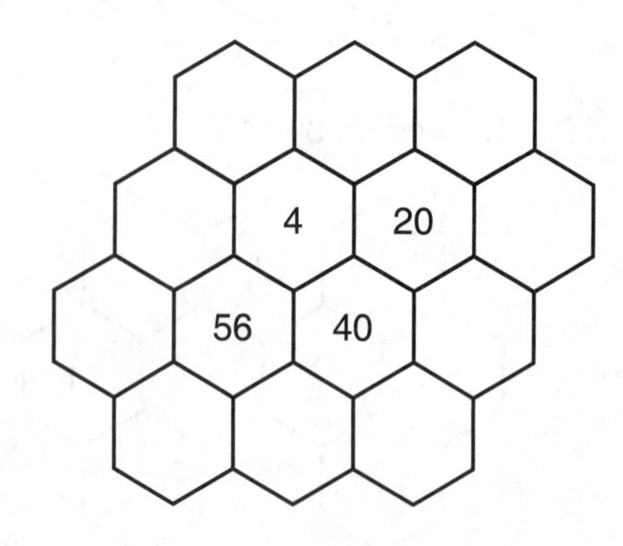

PRACTICE | Skip-count by the given number to solve each Honeycomb Path puzzle below.

177. ★ Skip-count by 5's.

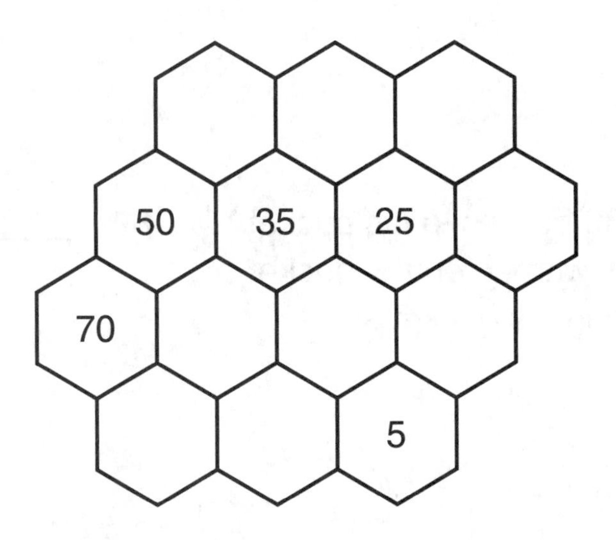

178. ★ Skip-count by 2's.

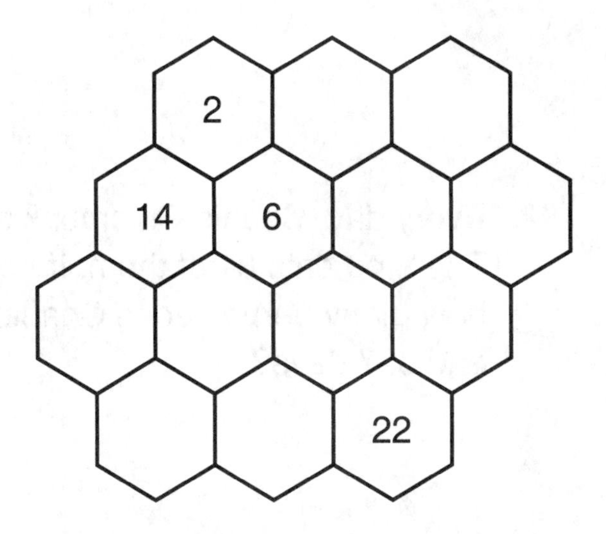

179. ★ Skip-count by 3's.

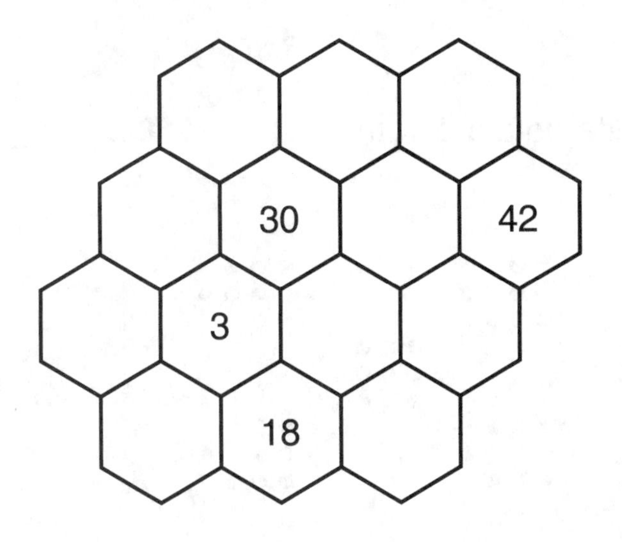

180. ★ Skip-count by 4's.

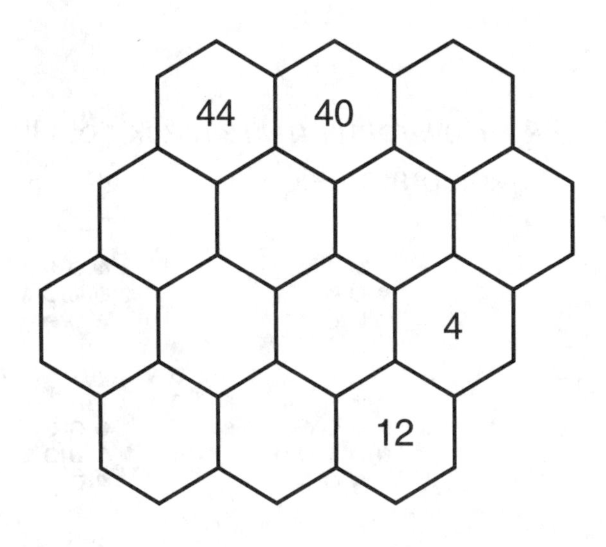

PRACTICE | Solve each problem below.

181. Compute $(777+444)-(777-444)$. **181.** ________

182. Every day, Charlie Chipmunk adds 36 berries to his bucket. Then, he eats 31 of them. If he starts with an empty bucket, how many berries does Charlie have in his bucket at the end of 7 days? **182.** ________

183. ★ Fill each circle below with + or − to make the equation true.

$$198 \bigcirc 55 \bigcirc 56 \bigcirc 57 \bigcirc 58 = 200$$

184. ★ How many ***more*** black dots than white dots are in the diagram below? **184.** ________

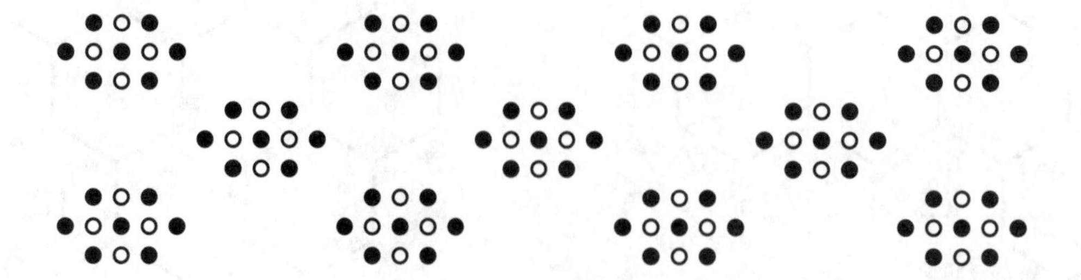

PRACTICE | Solve each problem below.

185. Fill in the blanks in the skip-counting pattern below.

_____, _____, 45, _____, _____, _____, _____, 60, _____, _____.

186. ★ Fill all three blanks below with the ***same number*** to make the equation true.

$44 + \_\_\_\_ - \_\_\_\_ + \_\_\_\_ = 100$

187. ★ Ralph adds every whole number from 1 to 99.
Cammie adds every whole number from 2 to 100.
How much greater is Cammie's sum than Ralph's sum?

187. 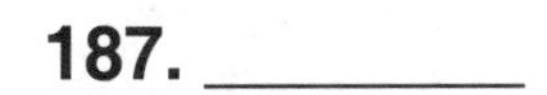

188. ★ Compute the value of the expression below.

188. 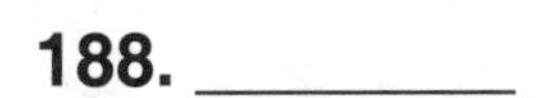

$(20+19)-(19+18)+(18+17)-(17+16)+(16+15)-(15+14)$

CHAPTER 9
Odds & Evens

Use this Practice book with Guide 2C from BeastAcademy.com.

Recommended Sequence:

Book	**Pages:**
Guide:	72-77
Practice:	71-81
Guide:	78-85
Practice:	82-87
Guide:	86-103
Practice:	88-101

You may also read the entire chapter in the Guide before beginning the Practice chapter.

The numbers we get when we count by 2's starting with zero are ***even***:

0, 2, 4, 6, 8, 10, 12, 14, 16, 18, and so on.

The rest of the whole numbers are ***odd***:

1, 3, 5, 7, 9, 11, 13, 15, 17, 19, and so on.

Even numbers end in 0, 2, 4, 6, or 8.

Odd numbers end in 1, 3, 5, 7, or 9.

PRACTICE | Solve each problem below.

1. Circle every ***even*** number below.

23 26 30 41 55 78

2. Circle every ***odd*** number below.

229 442 531 577 790 983

3. What is the largest two-digit odd number? **3.** ________

4. What is the largest three-digit even number? **4.** ________

5. How many odd numbers are between 46 and 64? **5.** ________

We can split every even number into groups of 2, or into 2 equal groups.

For example, the 12 dots below can be split into 6 groups of 2, or into 2 groups of 6.

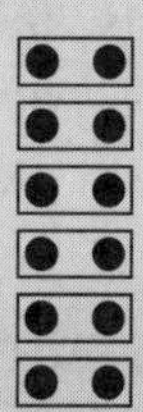 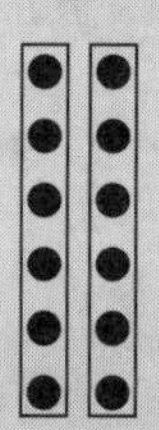

If we try to split an odd number into groups of 2, or into 2 equal groups, there will always be one extra.

For example, we can split 13 dots into 6 groups of 2, or into 2 groups of 6, with one extra.

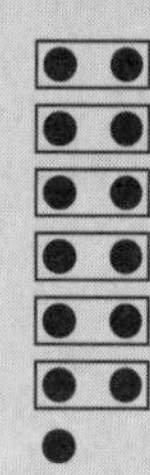 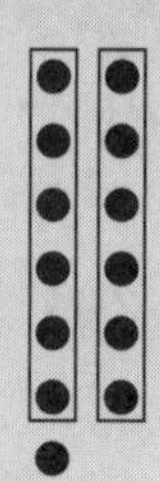

PRACTICE | Solve each problem below. Try to solve each problem ***without counting*** the number of dots.

6. Circle every figure below with an ***even*** number of dots.

 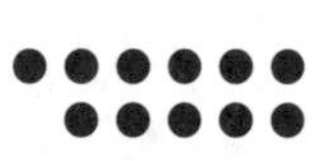 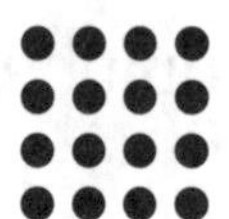  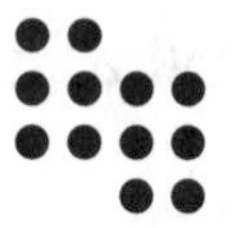

7. Circle every figure below with an ***odd*** number of dots.

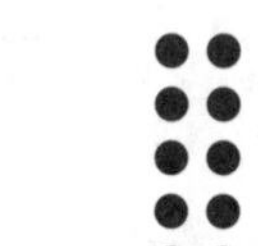 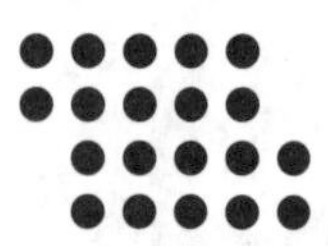 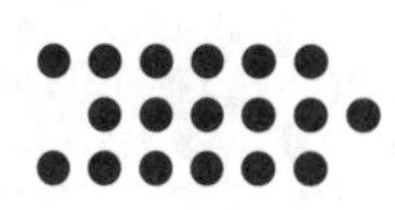 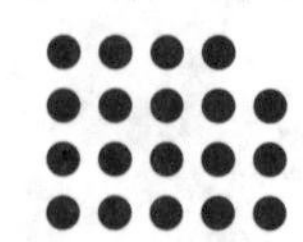 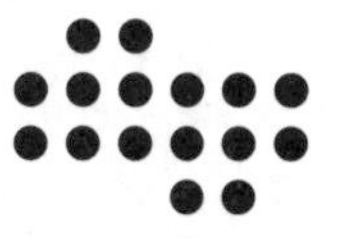

8. Circle every figure below with an ***even*** number of dots.

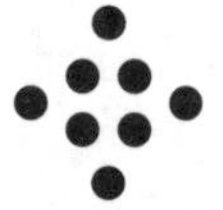 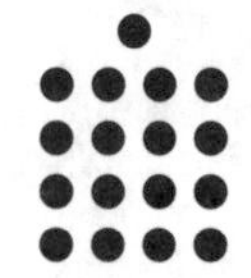 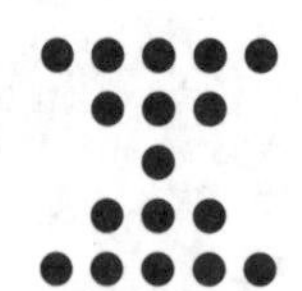 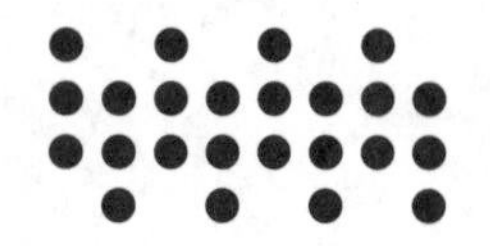 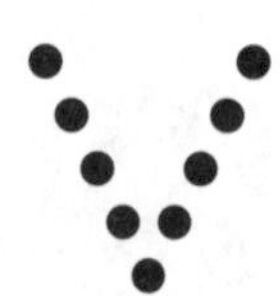

PRACTICE | To answer each question below, circle ***even*** or ***odd***.

9. Every jackalope in a field has two horns. Is the total number of horns on all of the jackalopes in the field even or odd?

9. even odd

10. Mergle invites monsters to her birthday party, and every monster she invites brings one friend. Including Mergle, is the total number of monsters at her party even or odd?

10. even odd

11. Rhea has a stack of index cards. She draws one shape on the front and back of each index card. Is the total number of shapes Rhea draws even or odd?

11. even odd

12. Beastball teams always have the same number of players. On the Beastball field, there are two teams and a referee. Is the total number of beasts on the field even or odd?

12. even odd

13. ★ A group of monsters high-five each other. Winnie asks each monster how many high fives they gave and adds their answers up. Is the total even or odd?

13. even odd

When we split a number into two equal amounts, each amount is ***half*** of the original number.

EXAMPLE | What is half of 16?

We can split 16 into two equal groups of 8.

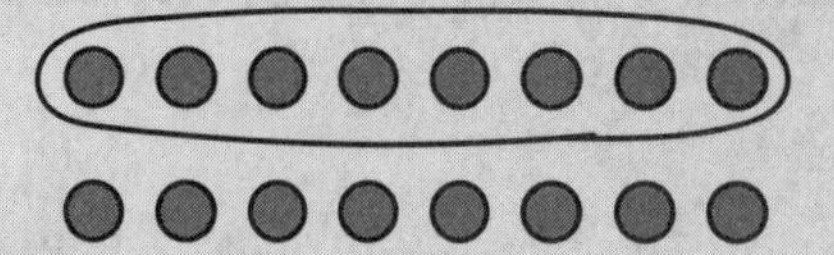

So, half of 16 is **8**.

– or –

We add two copies of 8 to get 16:
8 + 8 = 16. So, half of 16 is **8**.

PRACTICE | Solve each problem below.

14. Half of 26 is _______.

15. Half of 48 is _______.

16. Half of 406 is _______.

17. Half of 268 is _______.

18. Half of 34 is _______.

19. Half of 52 is _______.

20. Half of 250 is _______.

21. ★ Half of 770 is _______.

22. What number will you get if you double 479, then find half of the result?

22. _______

23. ★ What number is half of half of half of 144?

23. _______

PRACTICE | Answer each question below.

24. Vike has 147 marbles. Ash has 93 marbles. They decide to put their marbles into one pile and then share the pile equally. How many marbles will each monster get?

24. ________

25. Half of the students at Parity Elementary School are boys. Half of the boys at Parity Elementary School wear glasses. 48 boys wear glasses at Parity Elementary School. How many students attend the school?

25. ________

26. ★ Exactly half of the animals on Mr. Piggin's farm are chickens. Circle every number below that ***could*** be the number of animals on Mr. Piggin's farm.

18 25 76 98 243 777 952

27. ★ Each number in the list below is 10 more than ***half*** the number to its left. Fill in the blanks with the correct numbers to complete the list.

_____, _____, 52, 36, _____, _____, 22, _____

EXAMPLE | Circle the sums below that have an even result.

19+5 19+12 19+14

19+11 19+37 19+60

We compute each sum, then circle the results that are even.

(19+5 = 24) 19+12 = 31 19+14 = 33

(19+11 = 30) (19+37 = 56) 19+60 = 79

Notice that when we add an odd number to 19, the sum is always even. When we add an even number to 19, the sum is always odd.

PRACTICE | Solve each problem below.

28. Circle every sum below that has an ***even*** result.

8+13 8+21 8+24 8+42 8+53

29. Circle every sum below that has an ***even*** result.

23+6 23+12 23+34 23+53 23+89

30. Circle every sum below that has an ***odd*** result.

14+11 21+11 28+11 33+11 45+11

31. Circle every sum below that has an ***odd*** result.

19+14 32+14 38+14 55+14 91+14

PRACTICE | Write ***even*** or ***odd*** in each blank below.

32. Use the diagram to help you fill in the blank below.

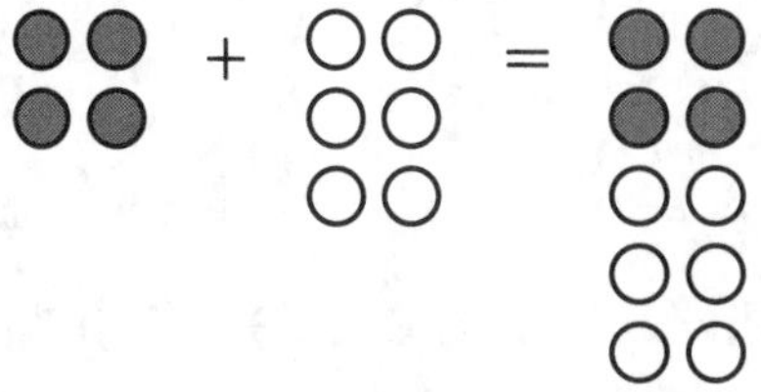

An even plus an even is always __________.

33. Use the diagram to help you fill in the blank below.

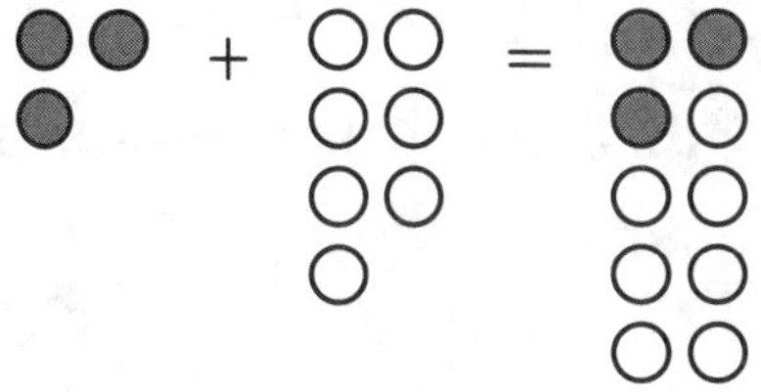

An odd plus an odd is always __________.

34. Use the diagram to help you fill in the blank below.

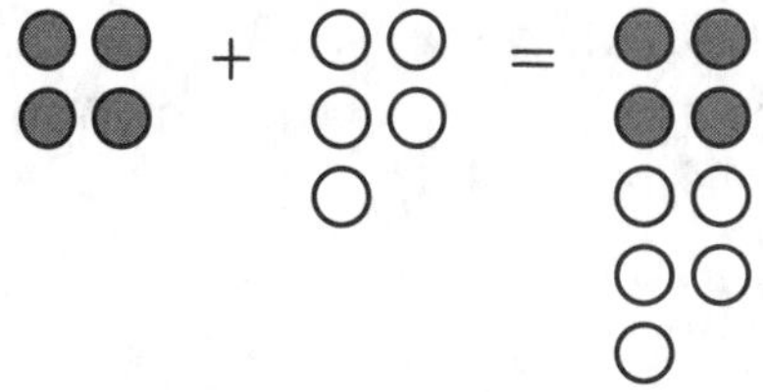

An even plus an odd is always __________.

35. Use the diagram to help you fill in the blank below.

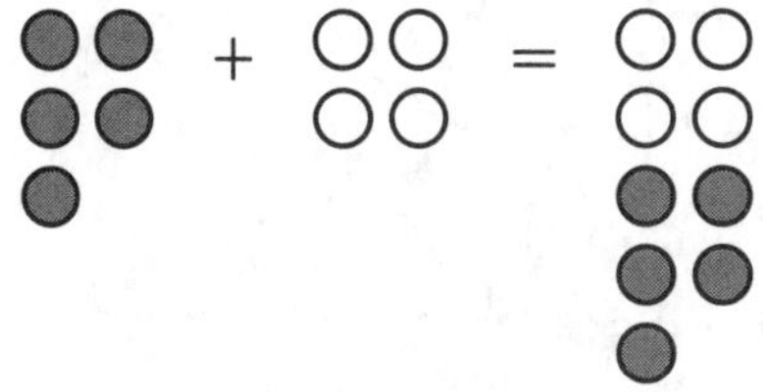

An odd plus an even is always __________.

PRACTICE | Solve each problem below.

36. Circle every sum below that has an ***even*** result.

419+853 824+786 742+217 819+777 203+698

37. Circle every sum below that has an ***odd*** result.

348+672 709+583 722+313 639+920 253+129

38. Circle every number below that is 244 more than an ***even*** number.

408 793 737 387 992

39. Circle every number below that is 159 more than an ***odd*** number.

331 284 931 646 445

40. ★ Place the digits 4, 6, 7, and 8 in the four small triangles below so that the sum of the digits in any two triangles that share a side is odd.

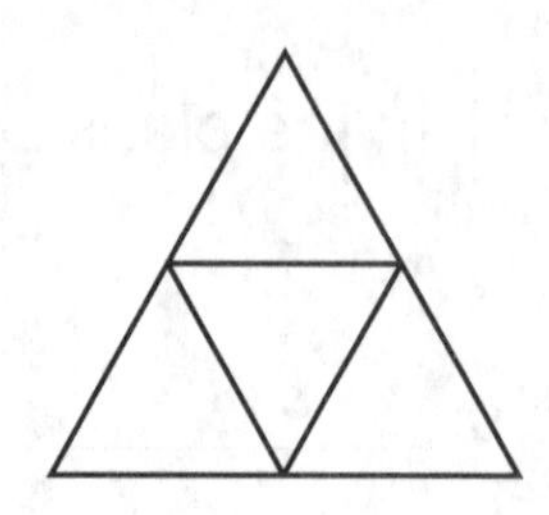

PRACTICE | Solve each problem below.

41. Five years ago, Finn's age in years was odd. Is Finn's age today even or odd?

41. ________

42. Abe tears a page out of his math workbook. He adds the page numbers from both sides. Is his result even or odd?

42. ________

43. Grogg and Lizzie have the same number of pencils. Circle every number below that could be the number of pencils Grogg and Lizzie have together.

25 17 34 56 41 97

44. ★ Alex has 5 more pencils than Winnie. Circle every number below that could be the number of pencils Alex and Winnie have together.

40 55 24 13 29 81

45. ★★ The digit-sum of a number is the sum of that number's digits. For example, the digit-sum of 45 is $4+5=9$.

How many ***even two-digit*** numbers have an ***odd*** digit-sum?

45. ________

PRACTICE | Write ***even*** or ***odd*** in each blank below.

46. Use the diagram to help you fill in the blank below.

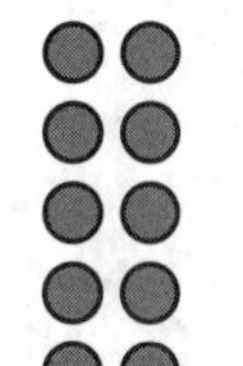
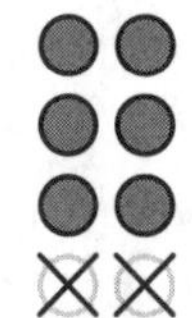

An even minus an even is always __________.

47. Use the diagram to help you fill in the blank below.

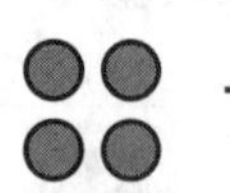
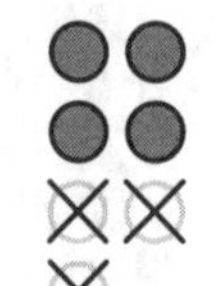

An odd minus an odd is always __________.

48. Use the diagram to help you fill in the blank below.

An even minus an odd is always __________.

49. Use the diagram to help you fill in the blank below.

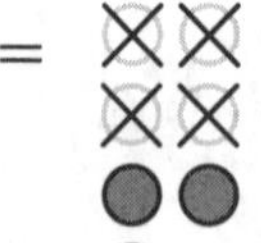

An odd minus an even is always __________.

PRACTICE | Solve each problem below.

50. Circle every difference below that has an ***odd*** result.

719 − 183 454 − 176 298 − 219 459 − 273 833 − 198

51. Circle every difference below that has an ***even*** result.

848 − 672 939 − 560 709 − 583 722 − 313 253 − 129

52. Grogg adds two numbers. Winnie finds the difference of the same two numbers. Circle the statement below that is ***impossible***.

Both results are odd. Both results are even. One result is odd, one result is even.

53. Marcus has two boxes of Cookie Critters. Both boxes hold the same number of cookies. Marcus eats 17 cookies. Is the number of cookies he has left odd or even?

53. ________

54. ★ Place the numbers 234, 456, 567, and 678 in the four small triangles below so that the difference between numbers in any two triangles that share a side is odd.

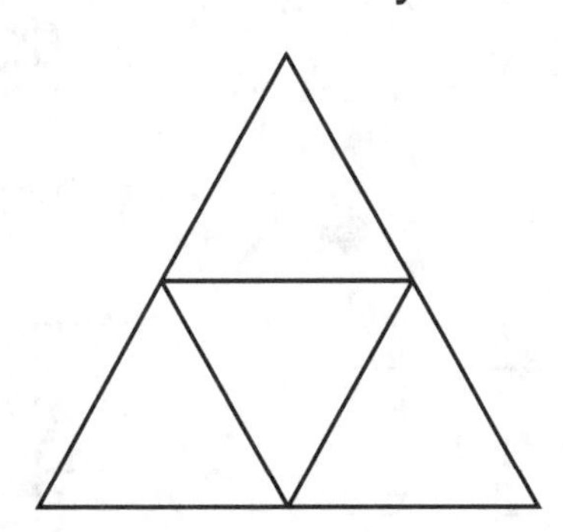

ODDS & EVENS
More Than Two Numbers

EXAMPLE | Is 83+84+85+86 even or odd?

We can add the numbers in any order. Since odd+even = odd, 83+84 and 85+86 both give odd sums.

Then, odd+odd = even.
So, 83+84+85+86 is **even**.

$$83+84+85+86$$
$$= \text{odd} + \text{odd}$$
$$= \text{even}$$

We can use what we know about adding two numbers to figure out whether the sum of more than two numbers is even or odd.

PRACTICE | Write ***even*** or ***odd*** in the blanks below to describe each sum.

55. 123+234+345 is ________ (even/odd).

56. 172+283+394 is ________ (even/odd).

57. 22+44+66+88 is ________ (even/odd).

58. 19+29+39+49+59 is ________ (even/odd).

59. 12+34+56+78+90 is ________ (even/odd).

60. 43+18+42+94+78 is ________ (even/odd).

61. The sum of 2 evens is _______.

The sum of 3 evens is _______.

The sum of 4 evens is _______.

The sum of 5 evens is _______.

The sum of 100 evens is _______.

The sum of 333 evens is _______.

62. The sum of 2 odds is _______.

The sum of 3 odds is _______.

The sum of 4 odds is _______.

The sum of 5 odds is _______.

The sum of 99 odds is _______.

The sum of 250 odds is _______.

PRACTICE | Solve each problem below.

63. Yuri's sticker book contains an odd number of pages, and each page contains an odd number of stickers. Is the total number of stickers in Yuri's book even or odd?

63. ________

64. All but one of the buses on Beast Island have an even number of wheels. Is the total number of wheels on all of the Beast Island buses even or odd?

64. ________

65. All of the monsters on a team have an even number of hands. Every hand has an odd number of fingers. Is the total number of fingers on all of the monsters even or odd?

65. ________

66. Sam adds an odd number of odd numbers, then subtracts an even number of even numbers from the result. Is Sam's final result even or odd?

66. ________

67. ★ What is the largest ***odd*** sum you can get by adding ***four*** of the numbers below?

67. ________

2 5 12 20 25 35 50

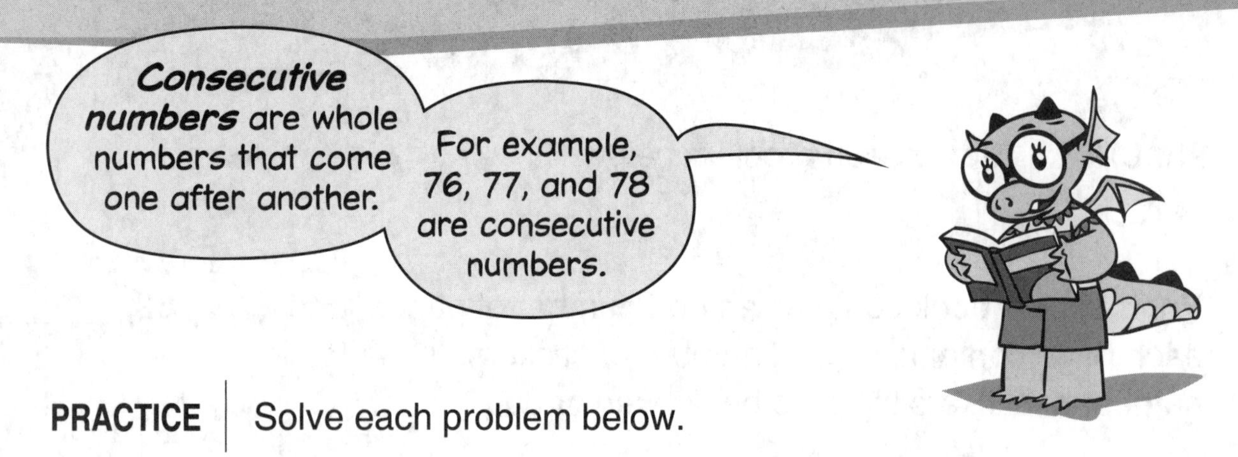

PRACTICE | Solve each problem below.

68. Can you find ***three consecutive numbers*** that have an ***odd*** sum?
If so, write all three numbers in the blank below. If not, write "Impossible."

____

69. Can you find ***three consecutive numbers*** that have an ***even*** sum?
If so, write all three numbers in the blank below. If not, write "Impossible."

____

70. Can you find ***four consecutive numbers*** that have an ***odd*** sum?
If so, write all four numbers in the blank below. If not, write "Impossible."

____

71. Can you find ***four consecutive numbers*** that have an ***even*** sum?
If so, write all four numbers in the blank below. If not, write "Impossible."

____

PRACTICE | Solve each problem below.

72. What is the largest ***odd*** result you can get by adding three different 2-digit numbers?

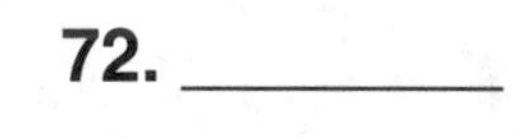

72. ________

73. Draw a line that separates the 7 numbers below into two groups so that each group has an even sum.

311 340 156 208 253 101 359

74. ★ Write an expression using the numbers 5, 4, 3, 2, and 1 once each with a + or − sign between each pair of numbers to get each result below. If it is not possible, write "Impossible" in the blank. Two of the blanks have been filled for you.

15 = 5+4+3+2+1 14 = ________

13 = ________ 12 = ________

11 = ________ 10 = ________

9 = ________ 8 = ________

7 = 5−4+3+2+1 6 = ________

5 = ________ 4 = ________

3 = ________ 2 = ________

1 = ________ 0 = ________

In an **8's and 9's** puzzle, we fill hexagons with 8's and 9's so that the sum of the numbers in every row is odd. Rows in this puzzle can be horizontal (↔) or diagonal (⤢ or ⤡).

EXAMPLE Fill the empty hexagons with 8's and 9's to solve the puzzle to the right.

We consider the shaded row of hexagons to the right. There are two 8's in this row. Since two evens plus an odd is odd, the empty hexagon in this row must be 9.

Next, we have two 9's in the row shown to the right. Since two odds plus an odd is odd, the empty hexagon in this row must be 9.

Next, we have one 9 and one empty hexagon in the row shown to the right. Since an odd plus an even is odd, the empty hexagon in this row must be 8.

We continue using the strategies described above to complete the puzzle as shown. We check to make sure that all rows in all directions have an odd sum.

PRACTICE In each puzzle below, fill the hexagons with 8's and 9's so that the sum of the numbers in every row is odd.

75.

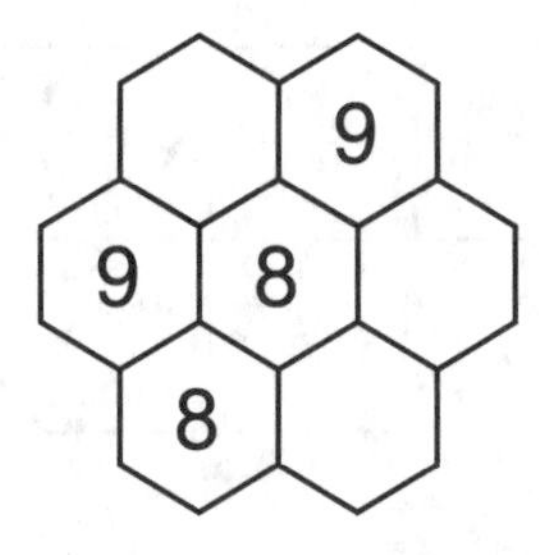

76.

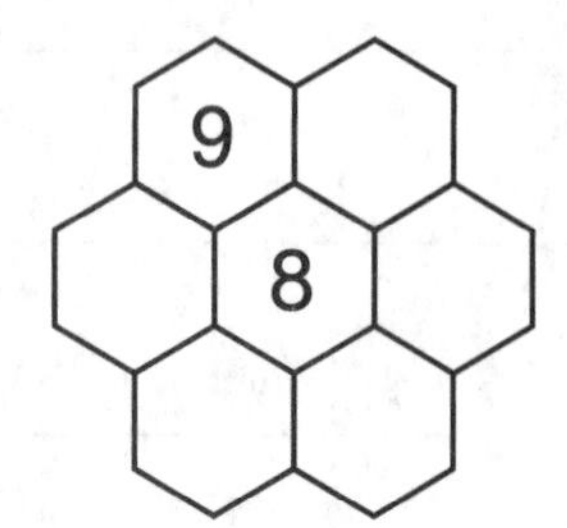

PRACTICE | In each puzzle below, fill the hexagons with 8's and 9's so that the sum of the numbers in every row is odd.

77.

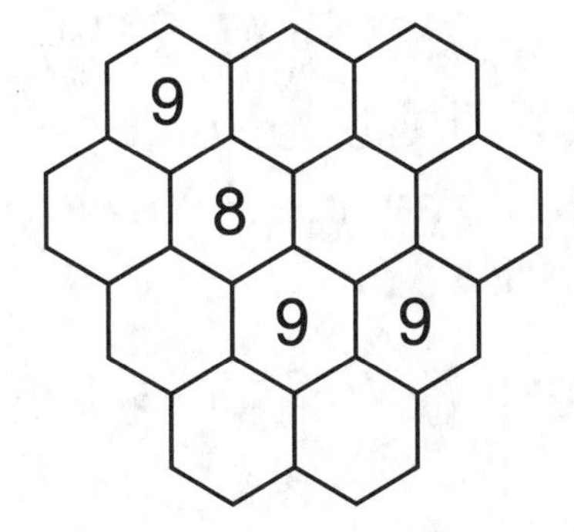

78.

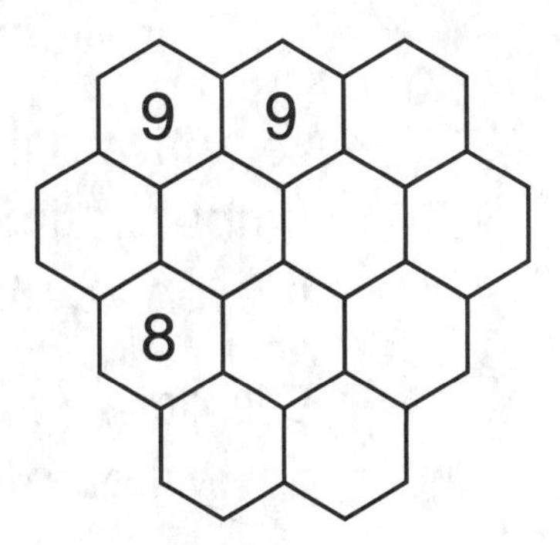

79.

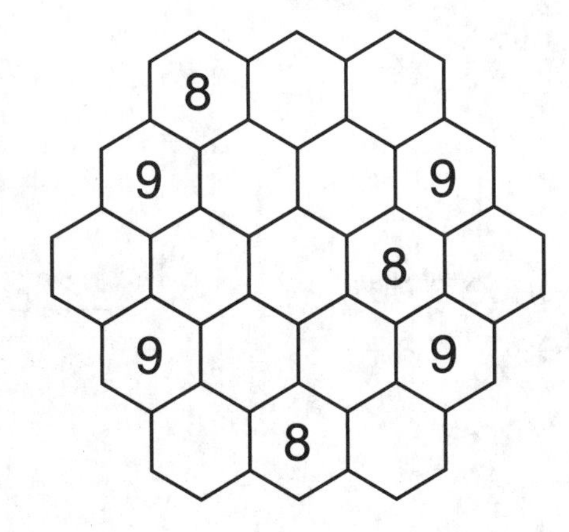

80.

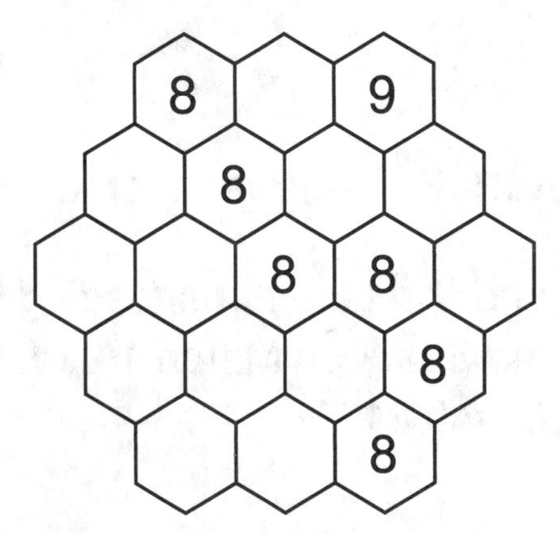

81.

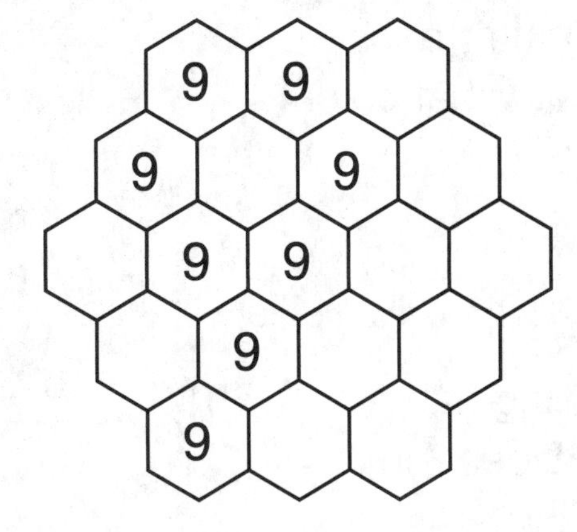

82. ★

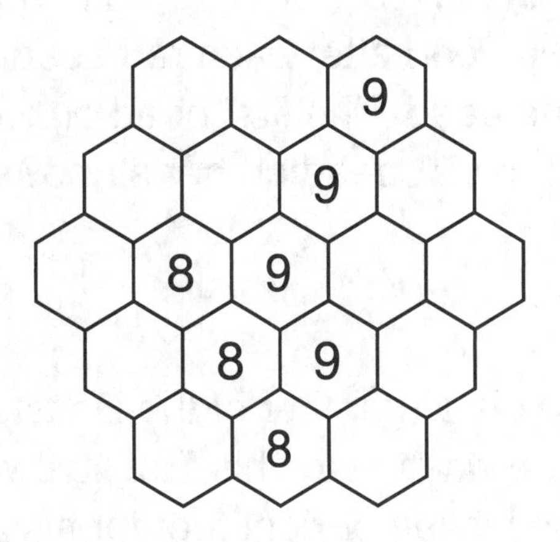

Print more of these puzzles at BeastAcademy.com.

EXAMPLE The light in Winnie's bedroom is currently on. If Winnie flips the light switch 99 times, will the light be on or off?

If Winnie flips the light switch 1 time, the light will be off.
If Winnie flips the light switch 2 times, the light will be on.
If Winnie flips the light switch 3 times, the light will be off.
If Winnie flips the light switch 4 times, the light will be on.

We notice a pattern! If Winnie flips the switch an odd number of times, the light will be off. If she flips the switch an even number of times, the light will be on.

99 is odd. So, if Winnie flips the light switch 99 times, the light will be **off**.

PRACTICE Answer each question below.

83. Ell and Mel take turns feeding their dog Spark once a day. Ell fed Spark on March 1st. Whose turn is it to feed Spark on March 31st?

83. ________

84. Ruth makes a string of red and blue beads. She places a blue bead after each red bead, and a red bead after each blue bead. The first bead on Ruth's string is red. What color is the 50th bead on her string?

84. ________

85. Alex is a member of the Beast Academy marching band. He always takes his first step with his left foot. Does Alex use his left or right foot for his 250th step?

85. ________

PRACTICE | Answer each question below.

86. Beast Grylls sets up camp next to a stream. Then, he crosses the stream 27 times while exploring. Will he need to cross the stream again to get back to his camp? Write ***yes*** or ***no***.

86. ________

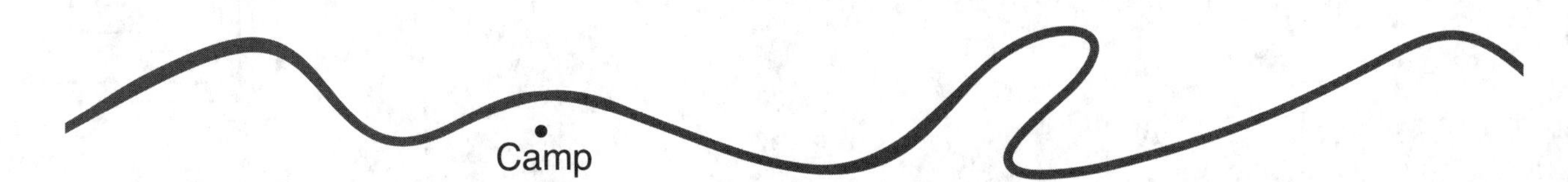

87. Grogg skip-counts by 3's, starting with 30. The first five numbers Grogg says are shown below.

87. ________

30, 33, 36, 39, 42

Is the 75th number Grogg says even or odd?

88. After ten minutes of Beastball play, the Crocadillos were ahead of the Elephrogs. Two minutes later, the Elephrogs took the lead. There were 15 ***more*** lead changes before the game ended. Which team won?

88. ______________

89. 175 little monsters line up boy-girl-boy-girl-boy-girl, and so on until the end of the line. The first monster in line is a boy. Circle the true statement.

There are more girls than boys in line.

There are more boys than girls in line.

There is the same number of girls as boys in line.

Every closed shape has an inside and an outside.

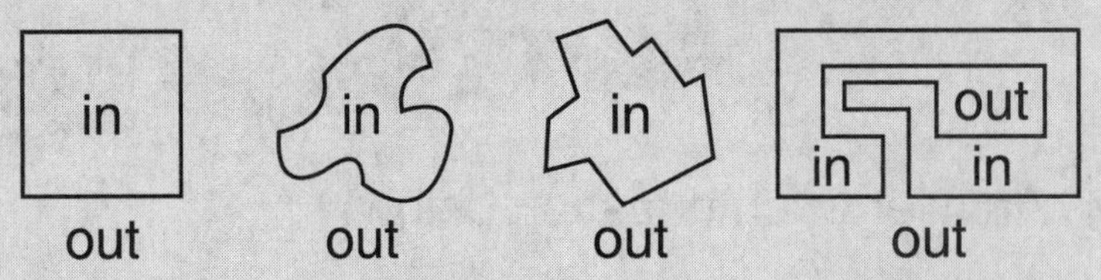

For complicated shapes, it's not always as easy to see which part is inside, and which part is outside.

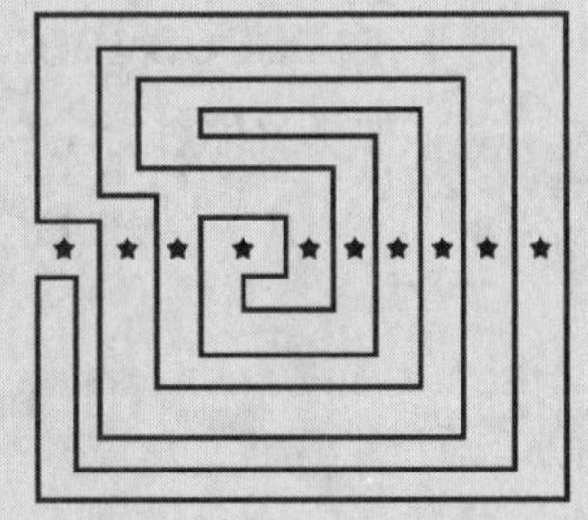

EXAMPLE | How many of the 10 stars are ***inside*** of the shape on the right?

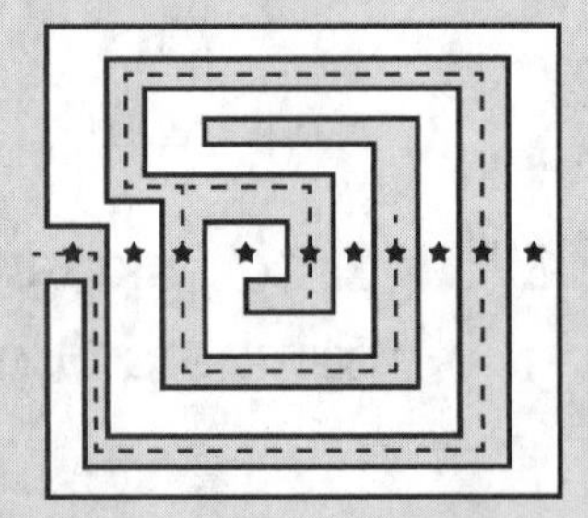

The 1st star on the left is outside of the shape. We can trace a path outside of the shape that crosses the 1st, 3rd, 5th, 7th, and 9th stars.

The **5** remaining stars are inside of the shape.

*Can you figure out which stars are outside of the shape **without** tracing a path or coloring the shape?*

PRACTICE | Solve each problem below.

90. Circle all of the stars that are ***inside*** of the shape below.

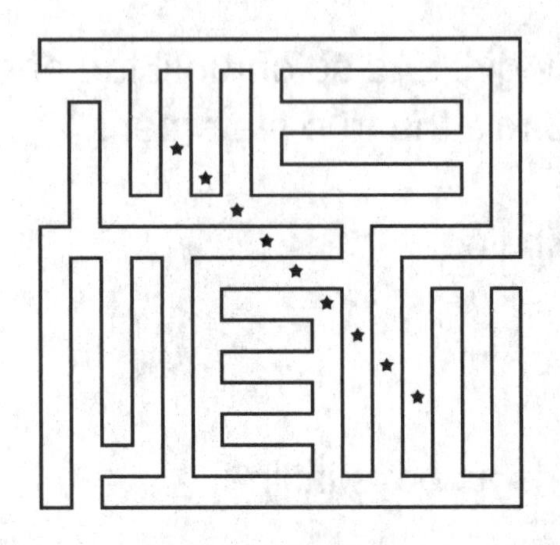

91. Circle all of the stars that are ***outside*** of the shape below.

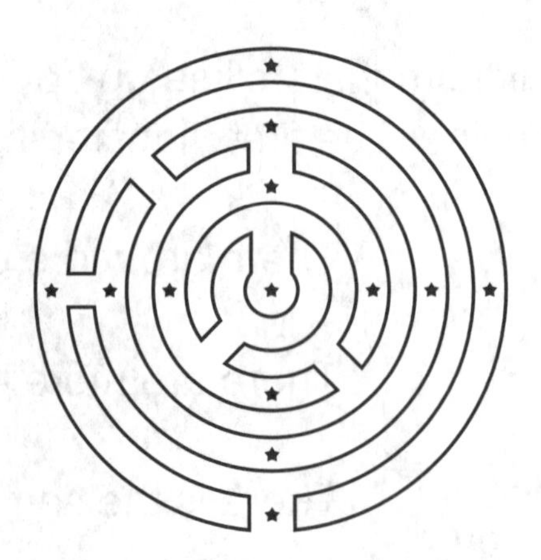

PRACTICE | Answer each question below about the shape on the right.

92. Label whether each letter listed below is ***inside*** or ***outside*** of the shape above.

A ________________ B ________________

C ________________ D ________________

93. Billy Bug is outside of the shape above. For each letter below, is the number of lines Billy must cross to get to it ***even*** or ***odd***?

A ________________ B ________________

C ________________ D ________________

94. Bobby Bug wants to walk between each pair of letters given below. For each pair, is the number of lines Bobby must cross ***even*** or ***odd***?

A ➡ B __________ A ➡ C __________

B ➡ C __________ B ➡ D __________

Every coin has two sides: heads and tails. We can turn a coin over from heads to tails, or from tails to heads.

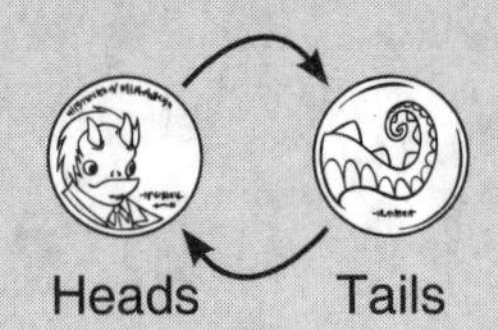

EXAMPLE Start with three heads. Turn over two coins at a time as many times as you want. Is it possible to get three tails?

Every time we turn over two coins, one of the following happens:

- We turn 2 heads to tails. This decreases the number of heads by 2.
- We turn 2 tails to heads. This increases the number of heads by 2.
- We turn 1 heads and 1 tails to 1 tails and 1 heads. This does not change the number of heads.

So, the number of heads can only go up or down by 2. We start with 3 heads, which is odd. Changing an odd by 2 gives an odd result. So, we can never get an even number of heads. Getting 0 heads (or 3 tails) is **impossible**!

PRACTICE For each arrangement below, turning ***any two*** coins is a "move."

Write the smallest number of moves needed to make all of the coins tails. If it is impossible, write "impossible."

95.      

95. ______________

96.    

96. ______________

97.      

97. ______________

98.         

98. ______________

PRACTICE

For each arrangement below, turning ***any three*** coins is a "move."

Write the smallest number of moves needed to make all of the coins tails. If it is impossible, write "impossible."

99.

99. ________________

100. ★

100. ________________

101. ★

101. ________________

PRACTICE

For each arrangement below, turning ***three coins that are next to each other*** is a "move."

Write the smallest number of moves needed to make all of the coins tails. If it is impossible, write "impossible."

102. ★

102. ________________

103. ★★

103. ________________

A ***checkerboard*** is a grid of squares in two colors.

A **Checkerboard Path** passes through every square on a checkerboard exactly once. Paths can only go up, down, left, or right (not diagonally).

EXAMPLE Trace a Checkerboard Path that begins in the upper-left square and ends in the bottom-right square of a 3-by-3 checkerboard.

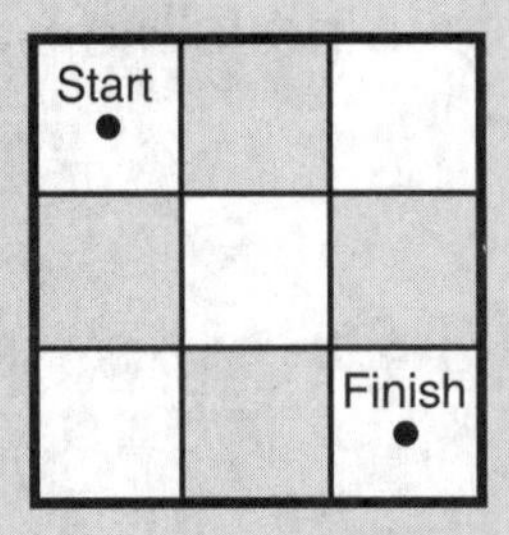

There are two different paths that pass through each square once before ending in the bottom-right square. Both paths are shown below.

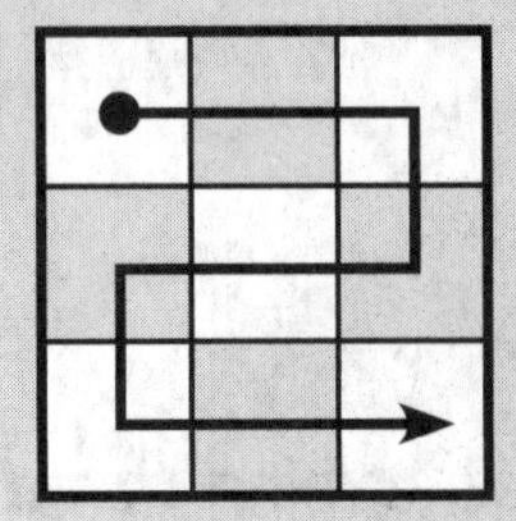

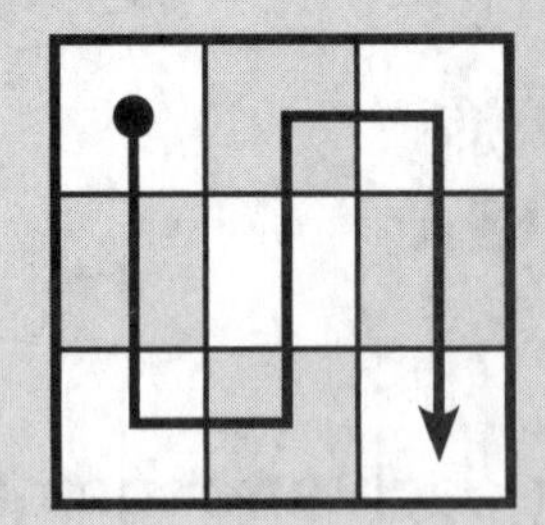

PRACTICE For each checkerboard below, trace a Checkerboard Path from Start to Finish that passes through each square exactly once. If it is ***impossible***, then circle the checkerboard.

104. Start Finish

105. Start Finish

106. Start Finish

PRACTICE For each checkerboard below, number the empty squares so that a Checkerboard Path can be traced through the squares in order from 1 to 16. If it is ***impossible***, then circle the checkerboard.

107.

1			
	9		
			16

108.

1	5		
			16

109.

1			16
	5		

110.

1			
	16	7	

111. ★

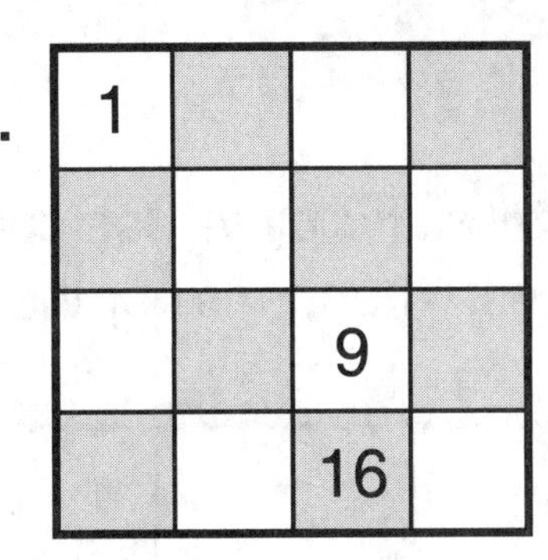

112. ★

1			
	8		

113. Circle the numbers below that appear in dark squares in the Checkerboard Paths above that are possible.

1 2 3 4 5 6 7 8 9 10 11 12 13 14 15 16

114. Circle the answer that completes the following sentence: For the Checkerboard Paths above that are possible, the even numbers appear in ____________.

dark squares light squares dark and light squares

115. Circle the boards below that are ***impossible*** to number.

1			
		16	

1			
		16	

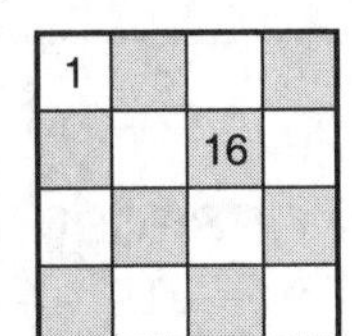

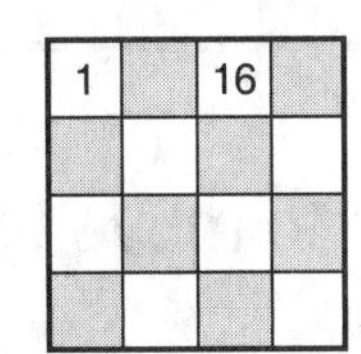

1		10	16

1	16		
			9

Fox & Rabbit is a game for two players. One player is the Fox, and the other is the Rabbit. The game is played on a grid of squares. Fox begins in the top-left square, and Rabbit begins in the bottom-right square.

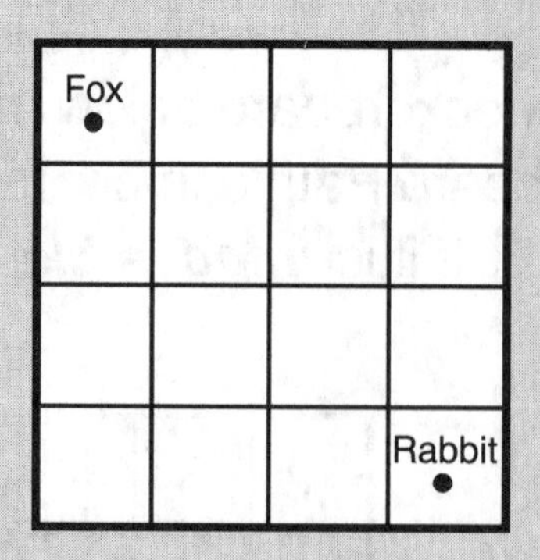

Fox and Rabbit take turns moving one square up, down, left, or right. If Fox catches Rabbit by landing on Rabbit's square, then Fox wins. If Fox cannot catch Rabbit, then Rabbit wins.

The player who is Rabbit chooses who moves first.

PRACTICE | Find a partner and play! Use coins or other small objects as game pieces. Then, answer the question that follows.

116. Should Rabbit choose to move first or second?

116. ______________

PRACTICE Find a partner and play on the 4-by-5 game board below. As before, Fox begins in the top-left square. Rabbit begins in the bottom-right square and gets to choose who moves first.

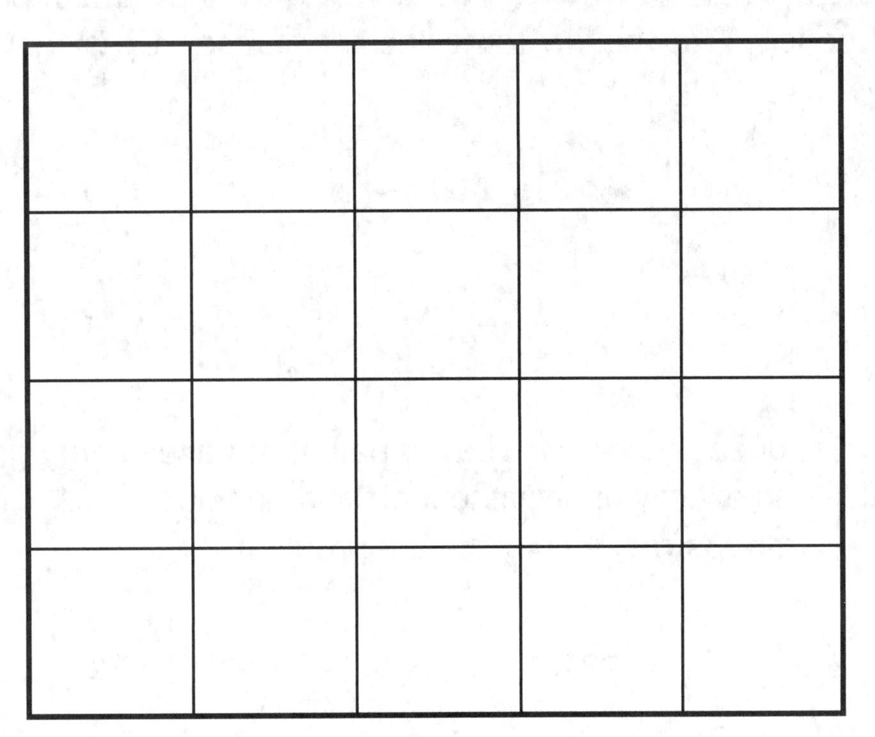

117. Color the squares like a checkerboard and play again. If Rabbit goes first, can Fox win?

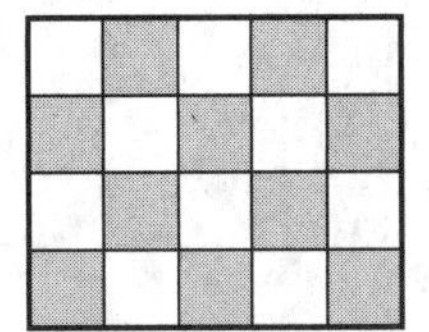

117. ______________

118. In a game of Fox & Rabbit played on a 5-by-5 board, should Rabbit choose to move first or second?

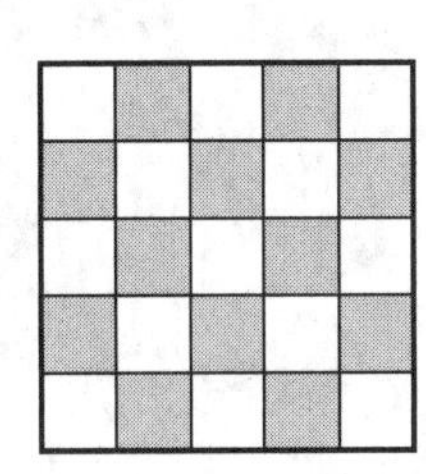

118. ______________

119. In a game of Fox and Rabbit, Fox is in a square next to Rabbit, but it is Rabbit's turn to move. Who will win the game?

119. ______________

Find printable game boards and variations at BeastAcademy.com.

EXAMPLE | Trace every gray line below without picking up your pencil or tracing over the same line twice.

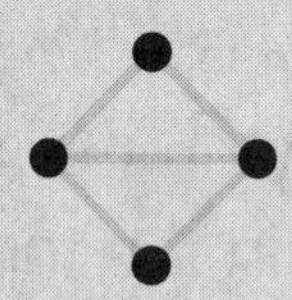

There are several ways to trace every gray line without picking up our pencil or tracing the same line twice. One way is shown below.

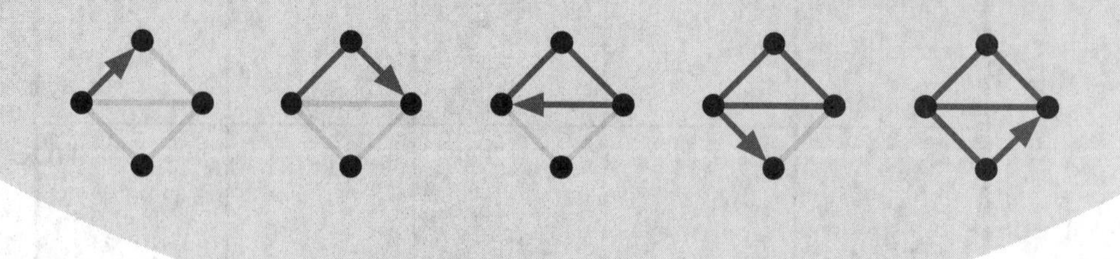

PRACTICE | For each problem, draw a path that traces every gray line exactly once without picking up your pencil. If it is ***impossible***, then circle the problem.

120.

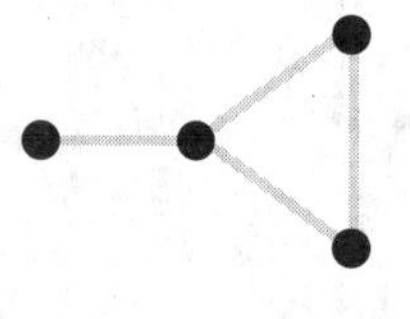

121.

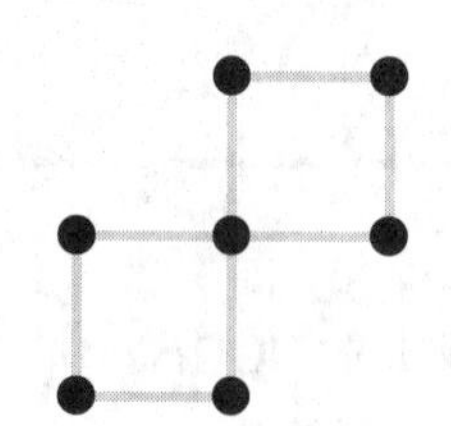

122.

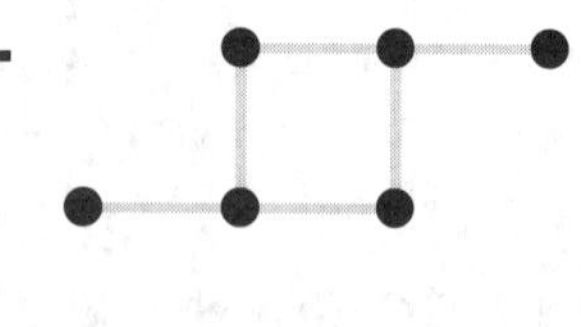

123.

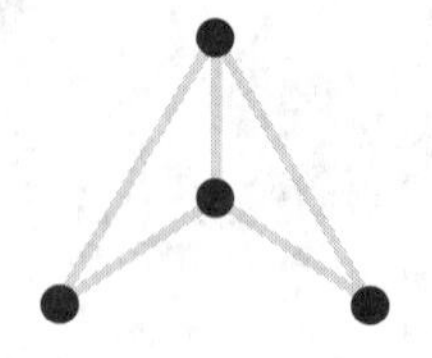

124.

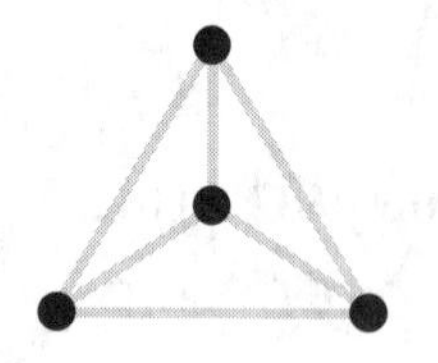

125.

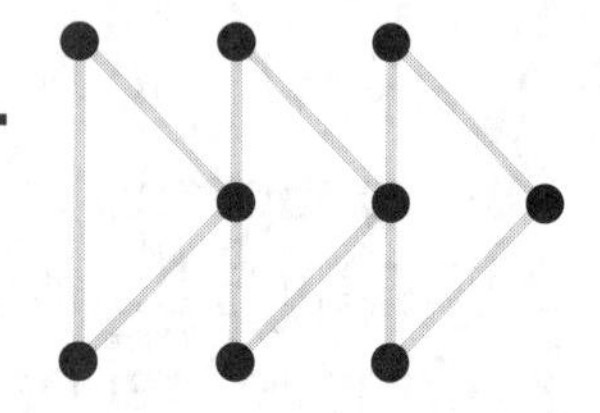

126.

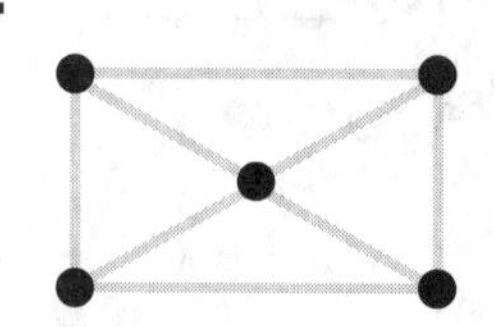

127.

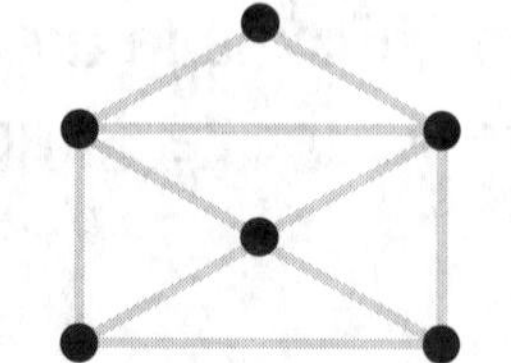

128.

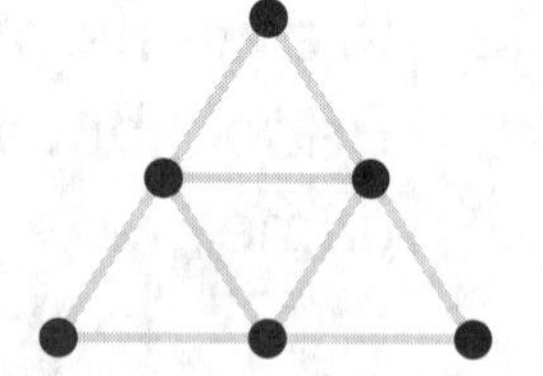

Print extra copies of these puzzles at BeastAcademy.com.

PRACTICE | Answer each question below.

129. We call a dot that is connected to an even number of dots an ***even dot***.
We call a dot that is connected to an odd number of dots an ***odd dot***.
On the previous page, circle every odd dot.

130. Fill each blank below with "possible" or "impossible" to describe the problems on the previous page.

If all the dots are even, it is ________________ to trace every line.

If there are exactly 2 odd dots, it is ________________ to trace every line.

If there are more than 2 odd dots, it is ________________ to trace every line.

131. Fill the blank below with "even" or "odd" to describe the problems on the previous page.

To trace a problem that has exactly 2 odd dots, we must start and end on an _________ dot.

132. ★ Circle the odd dots in each drawing below. Then, draw a path that traces every gray line exactly once without picking up your pencil.

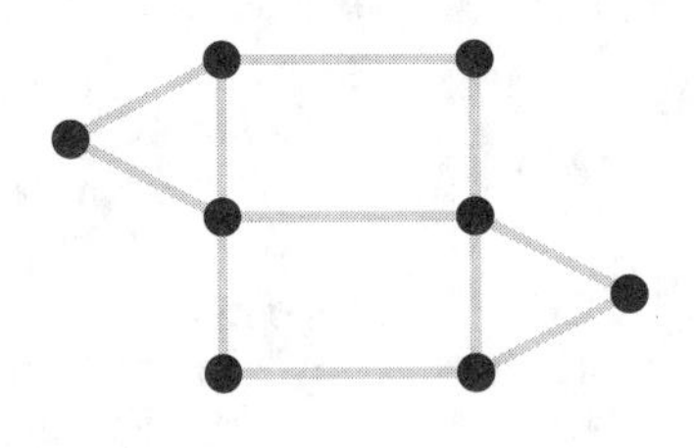

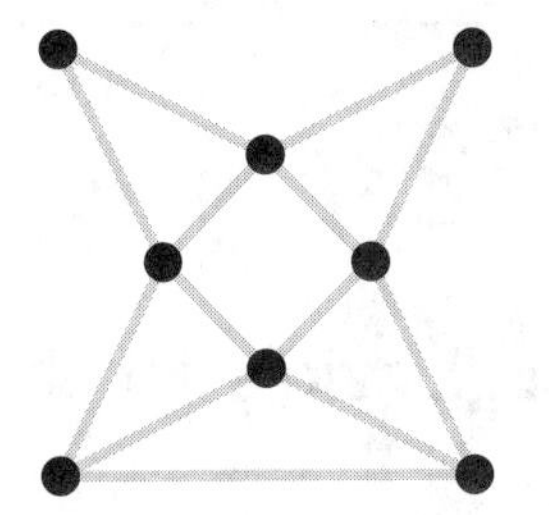

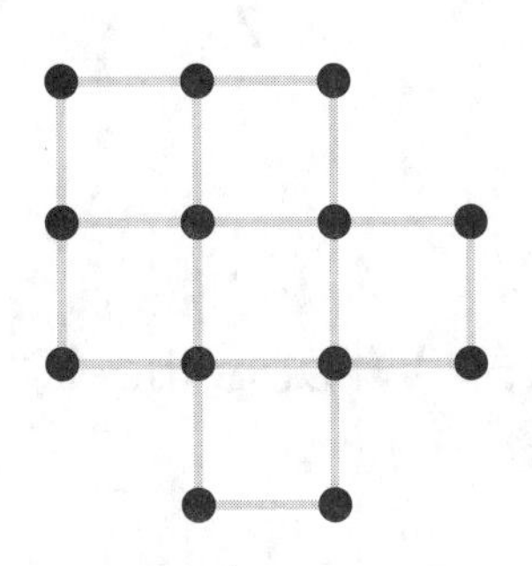

PRACTICE | Answer each question below.

133. ★ Rosa adds four consecutive whole numbers. Circle the only number below that could be her result.

33 44 55 66 77

134. ★ Draw two straight lines that split the 16 dots below into groups that all have a ***different*** odd number of dots.

● ● ● ●
● ● ● ●
● ● ● ●
● ● ● ●

135. Lizzie writes a list of numbers using the following rules:

- If a number is even, she takes half of it to get the next number.
- If a number is odd, she adds 3 to it to get the next number.

The first number in Lizzie's list is 57. Fill in the next eight numbers in her list.

57, _____, _____, _____, _____, _____, _____, _____

136. ★ What is the 100th number in Lizzie's list above? **136.** ________

PRACTICE | Answer each question below.

137. ★ Mike has between 30 and 40 pennies. He separates the pennies into two equal piles. Then, he separates each of these smaller piles into two equal piles. After this, Mike can no longer separate each pile into two equal piles. How many pennies does Mike have?

137. ________

138. ★ Draw a path that traces every gray line exactly once without picking up your pencil.

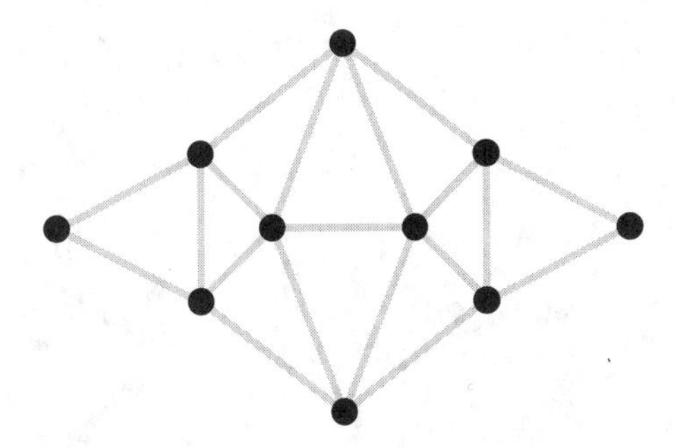

139. ★ Winnie writes 20 consecutive numbers from least to greatest. Exactly one of the statements below is true. Circle the statement that must be true.

Winnie's 5th number is odd.

Winnie's 6th number is odd.

Winnie's 7th number is odd.

Winnie's 8th number is even.

Winnie's 9th number is odd.

HINTS
For Selected Problems

HINTS
For Selected Problems

Below are hints to every problem marked with a ★.
Work on the problems for a while before looking at the hints.
The hint numbers match the problem numbers.

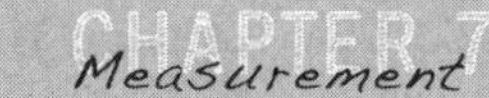

38. How many lines connect points that are 2 cm apart? 3 cm apart? 4 cm apart?

39. How many lines connect points that are 1 cm apart? 3 cm apart? 5 cm apart?

40. How many lines connect points that are 1 cm apart? 2 cm apart? 5 cm apart?

41. How many lines connect points that are 2 cm apart? 7 cm apart?

42. How many lines connect points that are 5 cm apart?

43. How many lines connect points that are 3 cm apart?

54. How many times must we add 4 inches to get 40 inches?

98. This pair of measurements must be connected as shown.

6 ft = 2 yd

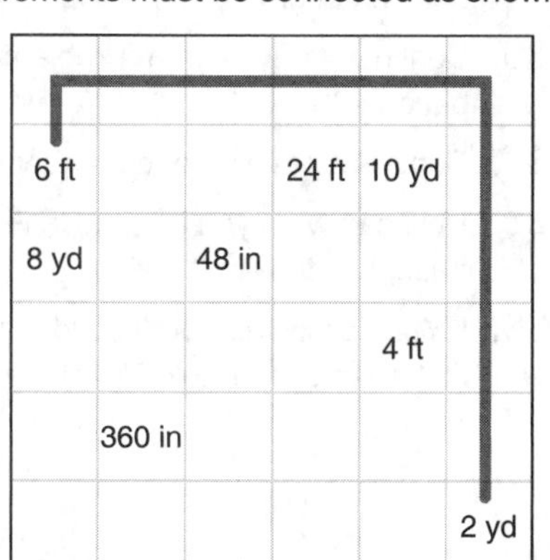

99. This pair of measurements must be connected as shown.

36 in = 3 ft

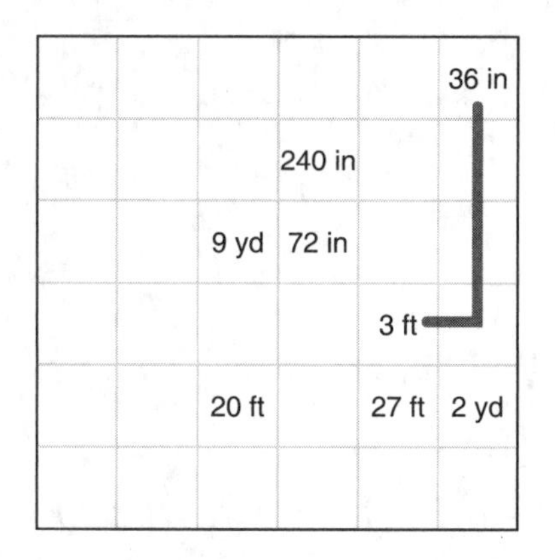

103. It may help to start by adding the feet and inches separately.

112. How many feet longer is 11 feet than *12* inches? How many inches is this?

113. How many feet is 4 yards?

114. What is the distance around the outside of the stop sign in feet? In inches?

— or —

If 8 sides of the stop sign have a total length of 2 yards, then what is the total length of 4 sides?

115. Keep organized. Connect the three plus signs at the bottom of the page. Then, connect each of these three points to the one in the top-left.

117. 35 inches is very close to how many feet?

— or —

What if Kirby traveled 1 inch with every hop? How many feet would he travel? What if he traveled 2 inches with every hop? 3 inches? 4 inches? 5 inches?

118. 5 feet is how many inches?

119. How many inches long is the short side of each small rectangle?

CHAPTER 8
Strategies 36-69

19. Can any numbers be paired to make easier sums?

57. Which two entries must be placed in each column?

58. Which two entries must be placed in each column?

59. Which two entries must be placed in each column?

60. Which two entries must be placed in each column?

61. Which two entries must be placed in each column?

62. Which two entries must be placed in each column?

86. What pairs of numbers in this expression almost cancel?

101. Can you simplify the equation before filling the blank?

102. We start with 77 and end up with 77. What must be true about +144, −12, +☐, and −145?

103. Starting with 200, we add and subtract some numbers to get 203. What does that tell us about the numbers we add and subtract?

112. How many of the +7 squares must the path cross?

113. Is it possible to get from 50 to 100 without crossing the +50 square?

114. How can we trace a path that gives a result with ones digit 4?

115. What is the smallest result you can get by only crossing addition squares?

116. How many of the +20 squares must the path cross?

117. How many squares that end in 1 must the path cross?

140. Only two expressions in parentheses cancel. Which two?

154. What is an easy way to add 99?

175. How can we connect 200 to 250? Then, how can we connect 250 to 325?

176. 56 is the largest number in this Honeycomb Path. Where must 52 go?

177. How can we connect 50 to 70?

178. How can we connect 6 to 14 without cutting off the path from 14 to 22?

179. Can 6 go in any of the hexagons to the left of 3?

180. Where must 8 go? Then, how many hexagons connect 12 to 40?

183. We start with 198 and end up with 200. What does that tell us about the numbers we add and subtract?

184. How could you solve this problem *without* finding the total numbers of black dots and white dots?

185. This pattern does not begin with the number we are skip-counting by. What numbers can you skip-count by to get from 45 to 60? Which of these uses the number of steps given?

186. Can this equation be simplified?

187. How much more is Cammie's 1st number than Ralph's? How much more is Cammie's 2nd number than Ralph's? How much more is Cammie's 3rd number than Ralph's?

— or —

Many of the numbers that Ralph and Cammie add are the same. What numbers are different?

188. What is (20+19)−(19+18)?

CHAPTER 9

Odds & Evens 70-101

13. Every high-five takes place between two monsters.
So, every high-five gets counted by both of the monsters who high-fived each other.

21. What is half of 700? Half of 70?

23. What's half of 144?
What's half of half of 144?

26. If half of the animals are chickens, then the total number of animals is double the number of chickens.

27. 52 is 10 more than half the number to its left. What is half of the number to the left of 52?

40. What numbers *cannot* share a side with 4? 6? 7? 8?

44. If Alex and Winnie had the *same* number of pencils, would the total number of pencils be even or odd?

45. What ones digits can be used?
What tens digits can be used?

54. What numbers cannot share a side with 234? 456? 567? 678?

67. How many of the 4 numbers could be odd?

74. Start with 15=5+4+3+2+1. What happens to the sum when you change a + to a −?

82. If you get stuck, you may have to guess whether a hexagon is 8 or 9. Try both and see what happens. There is only one correct answer!

100. To get all tails, you first need to get 2 tails and 3 heads. Can you do it?

101. To get all tails, you first need to get 1 tails and 3 heads. Can you do it?

102. What happens to the two middle coins with every move?

103. To get all tails, you first need to get 2 tails that are next to each other and 3 heads that are next to each other. Can you do it?

111. What number must be placed in the bottom-right square?

112. What color square must the 2 go in? The 4? The 6?

132. How can what you learned in Problem 131 help you trace these drawings?

133. Is the sum of four consecutive whole numbers even or odd?

134. How many groups must there be? How many dots must be in each group?

136. Continue your list from Problem 135. Do you notice any patterns?

137. Can Mike have an odd number of pennies?

138. How can what you learned in Problems 120-132 help you trace this drawing?

139. If Winnie's 5th number is odd, how many of the statements are true? What if her 5th number is even?

SOLUTIONS

Chapters 7-9

Chapter 7: Solutions

Comparing 7–9

1. The line is slightly **shorter** than the Ace of Spades from a standard deck of cards.

2. The line is slightly **longer** than a AA battery.

3. The line is slightly **shorter** than the long edge of a U.S. dollar bill.

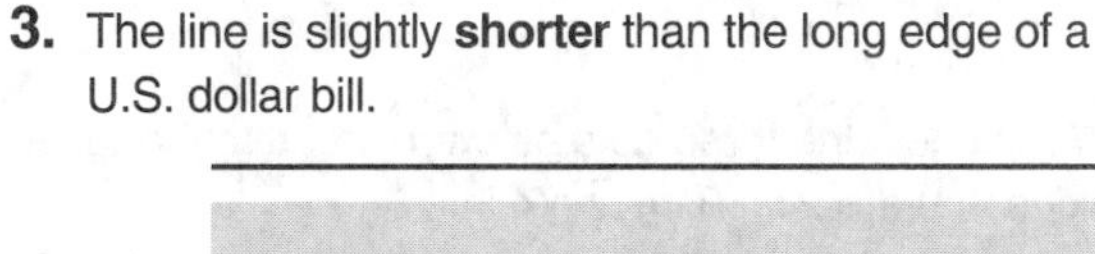

4. Printer paper is slightly **longer** than a page in this book.

5. The crayon on the right is longer than the crayon on the left.

6. The match on the left is longer than the match on the right.

7. The paper clip on the right is longer than the paper clip on the left.

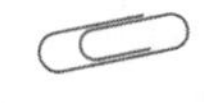 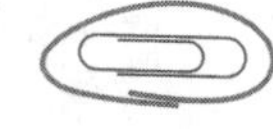

8. We mark the lengths of all three lines on a piece of paper, as shown below.

right line
left line
middle line

So, from shortest (1) to longest (3), we have:

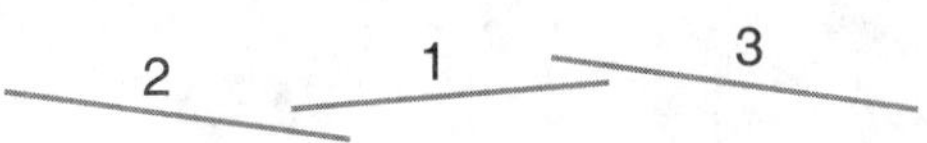

9. We mark the lengths of the fish as we did in Problem 8. From shortest (1) to longest (3), we have:

10. We mark the lengths of the shoes as we did in Problem 8. From shortest (1) to longest (3), we have:

  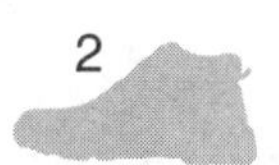

Units 10

11. The line is **3** dimes long.

12. The line is **8** dimes long.

13. The line is **6** dimes long.

14. The line is **9** dimes long.

15. The line is **7** dimes long.

MEASUREMENT

Using a Ruler 12-15

The rulers that appear in these solutions are not actual size.

16. The line is **5** inches long.

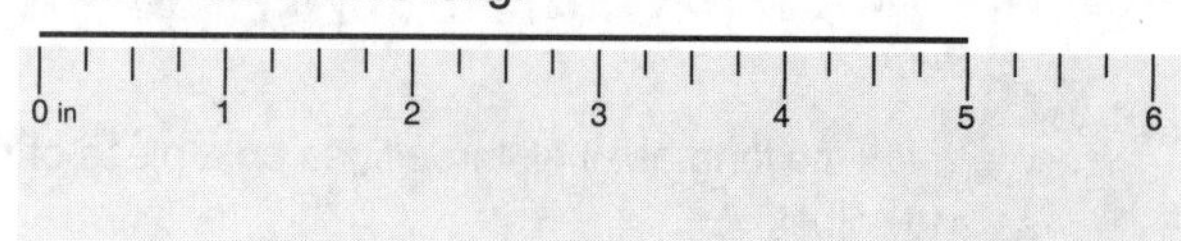

17. The line is **4** inches long.

18. The line is **6** inches long.

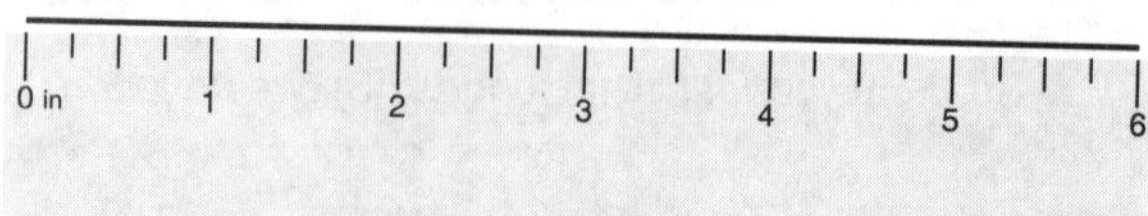

19. The line is **11** centimeters long.

20. The line is **9** centimeters long.

21. The sides are 5 cm, 12 cm, and 14 cm, as shown below.

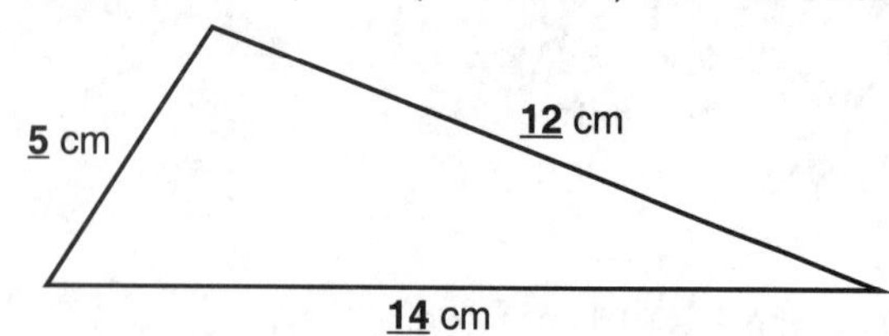

22. We connect the two dots and write "2 inches" on the line as shown below.

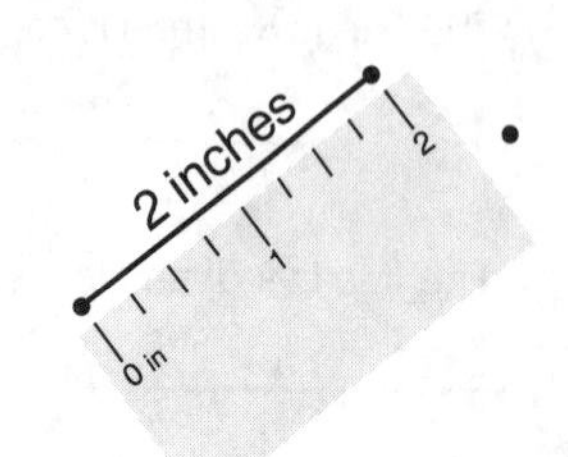

23. We connect the two dots and write "3 inches" on the line as shown below.

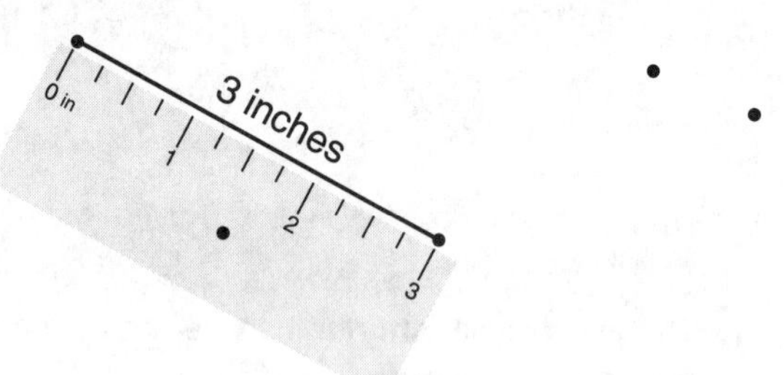

24. We connect the two dots and write "4 inches" on the line as shown below.

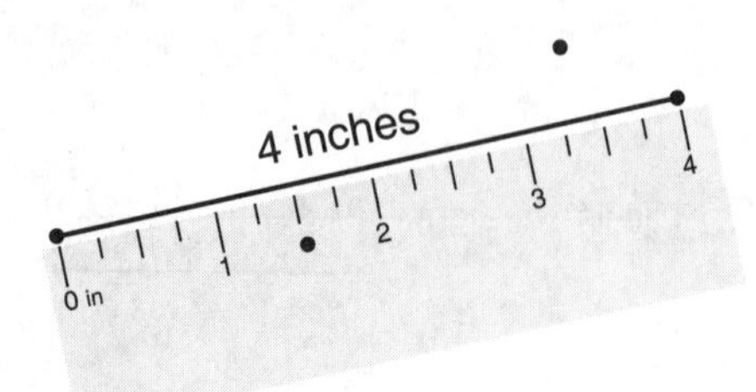

25. We connect the three dots below to create a triangle with sides that are 2 in, 3 in, and 4 in.

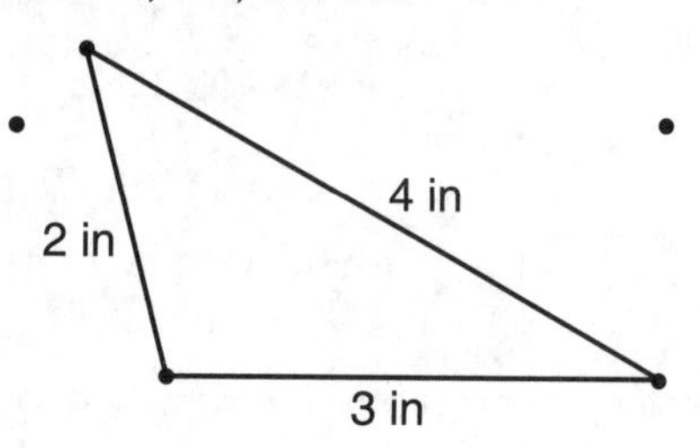

26. We place the 0 cm mark of the ruler at the dot marked F. We draw a dot on the line at the 10 cm mark.

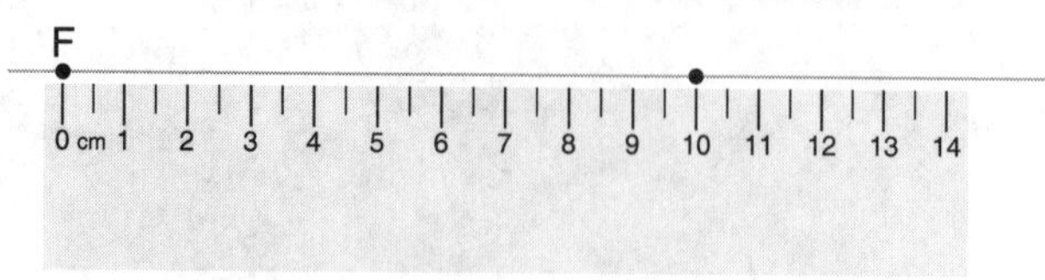

27. We place the 0 cm mark of the ruler at the dot marked G. We draw a dot on the line at the 11 cm mark.

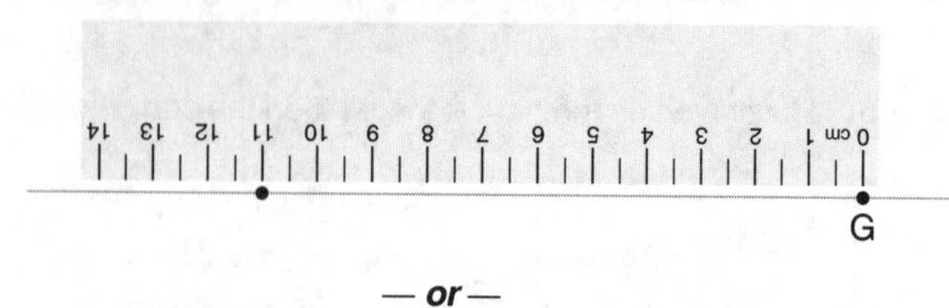

— or —

We place the 11 cm mark of the ruler at the dot marked G, then draw a dot on the line at the 0 cm mark.

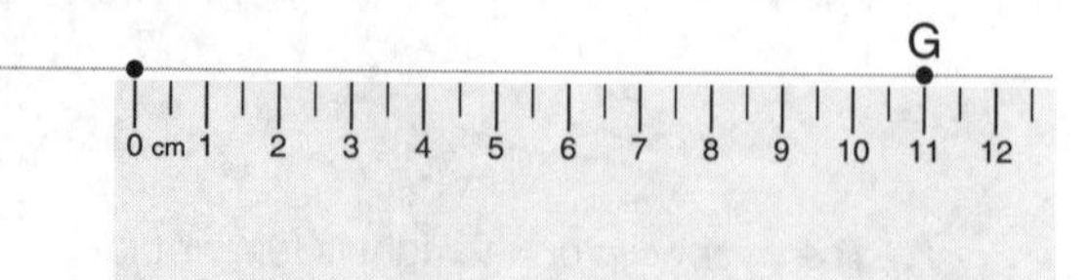

28. We place the 0 cm mark of the ruler at the dot marked H. We turn our ruler until the 12 cm mark lands on the line. We draw a dot on the line at the 12 cm mark.

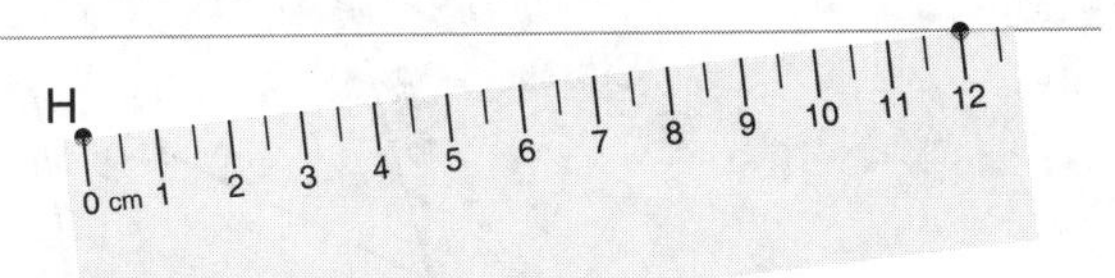

29. We place the 0 cm mark of the ruler at the dot marked J and turn our ruler until the 5 cm mark lands on the line. We can do this on the right and on the left. We draw dots on the line at the 5 cm mark on both sides of the J.

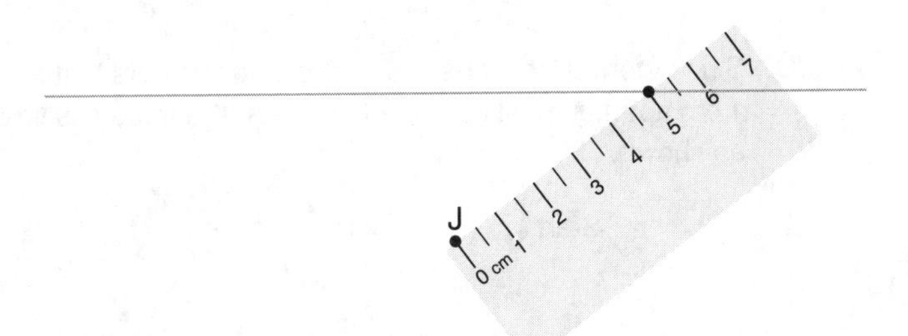

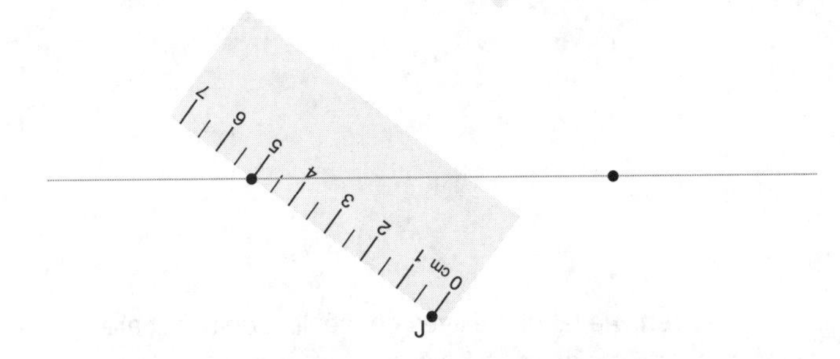

MEASUREMENT

Measure-Mazes

30. From the point labeled "Start," there is only one line connecting a point that is 1 cm away.

Then, there is only one line connecting a point that is 3 cm away.

Then, there is only one line connecting a point that is 4 cm away.

So, we connect the dots using the given distances in order as shown.

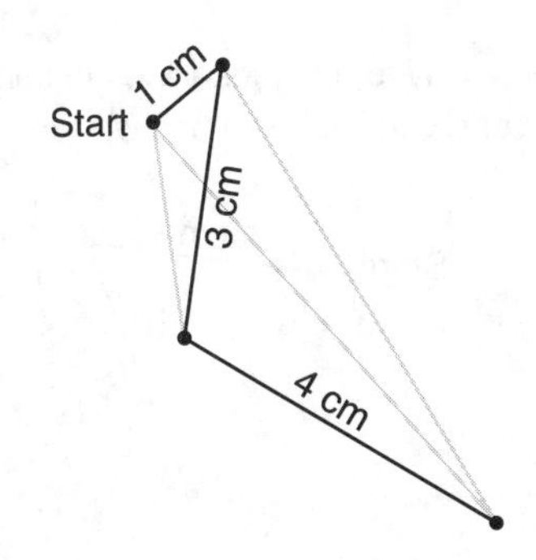

31. The only way to connect all of the dots using the given distances in order is shown below.

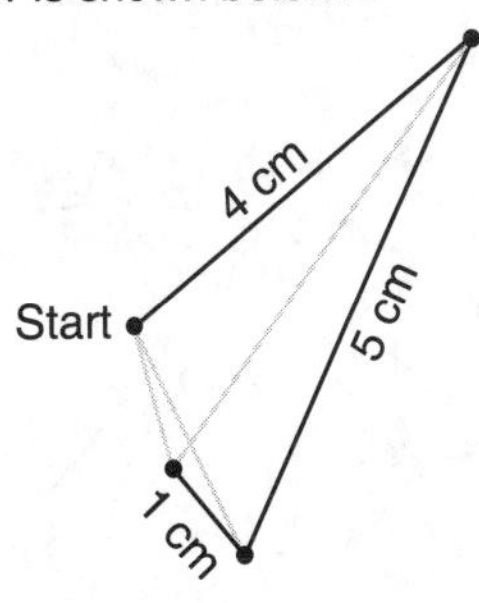

32. The only way to connect all of the dots using the given distances in order is shown below.

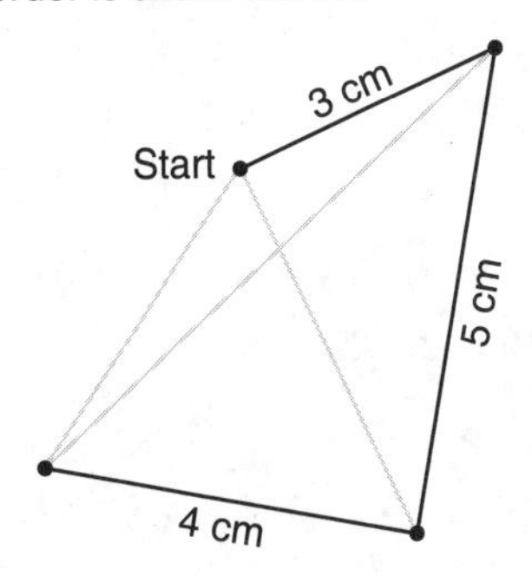

33. The only way to connect all of the dots using the given distances in order is shown below.

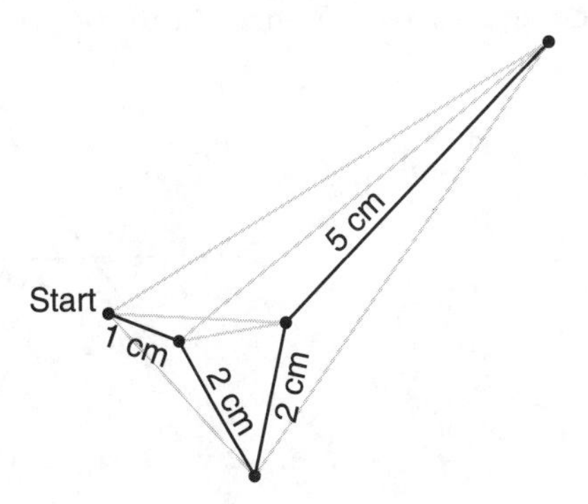

34. The only way to connect all of the dots using the given distances in order is shown below.

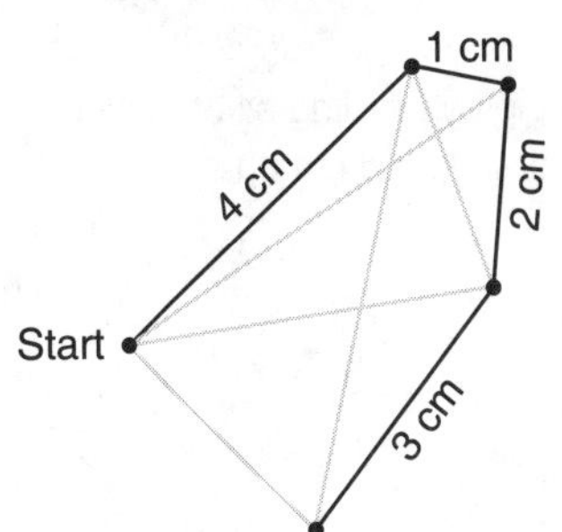

35. The only way to connect all of the dots using the given distances in order is shown below.

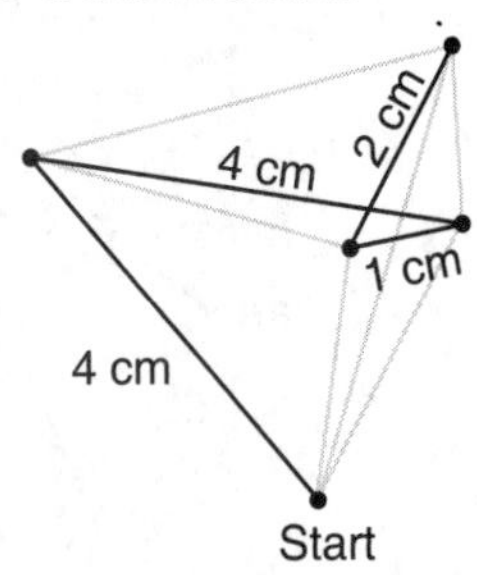

36. The only way to connect all of the dots using the given distances in order is shown below.

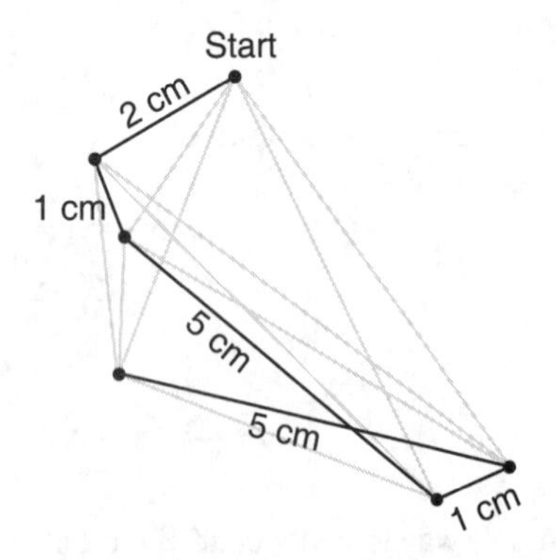

37. The only way to connect all of the dots using the given distances in order is shown below.

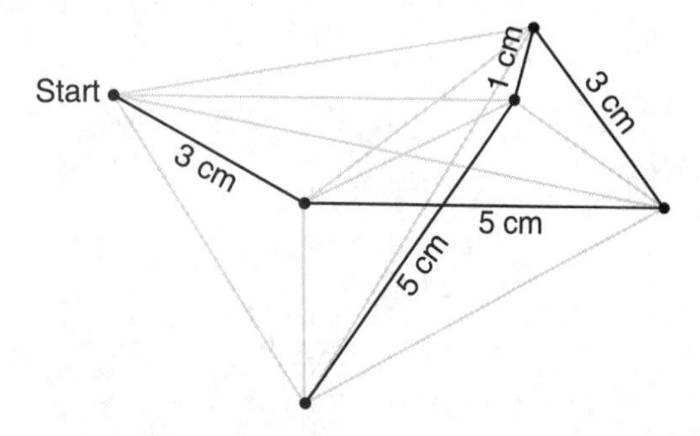

38. From the point labeled "Start," there are two different lines connecting a point that is 2 cm away. There are also multiple lines connecting points that are 3 cm and 4 cm apart.

We begin by labeling all of the lines connecting points that are 2, 3, or 4 cm apart.

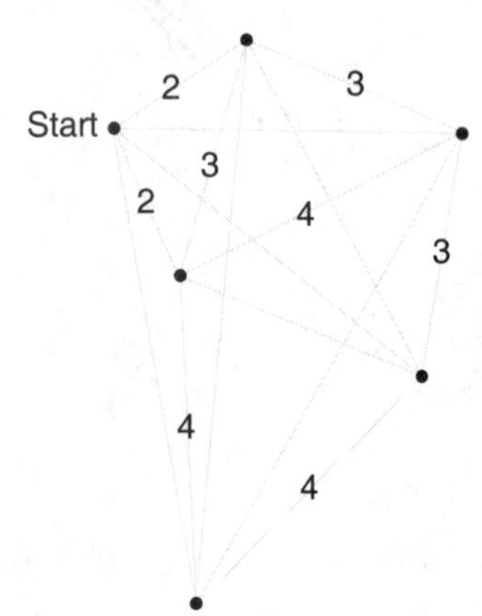

The only way to connect all of the dots using the given distances in order is shown below.

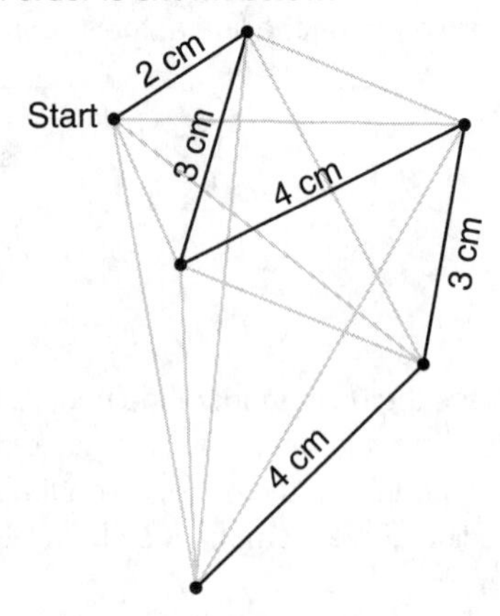

39. Since there is only one line connecting two dots that are 1 cm apart, it must be part of the path. We trace this line as shown.

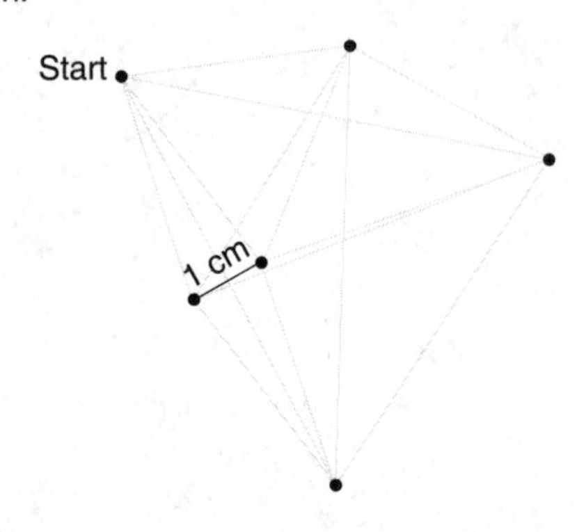

Then, we label the lines connecting dots that are 3 or 5 cm apart.

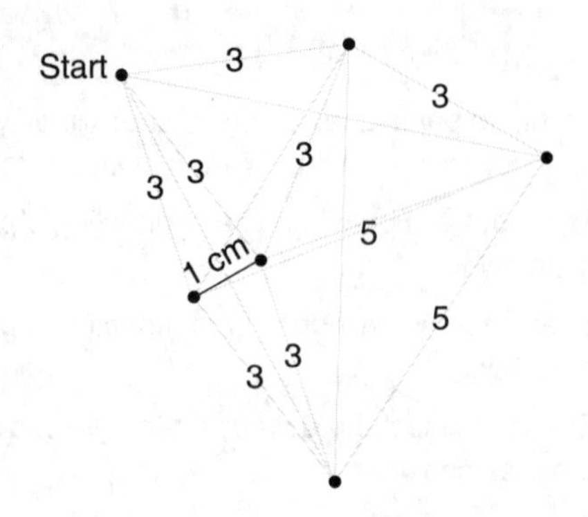

The only way to use these lines to complete the path in the order given is shown below.

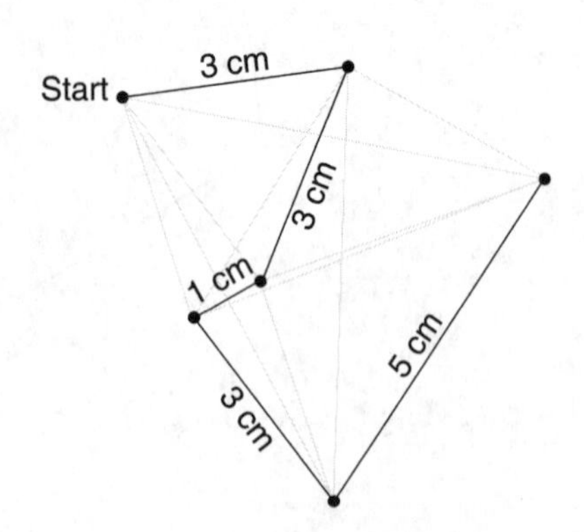

40. Since there is only one line connecting two dots that are 1 cm apart, it must be part of the path. We trace this line as shown.

Then, we label the lines connecting dots that are 2 or 5 cm apart.

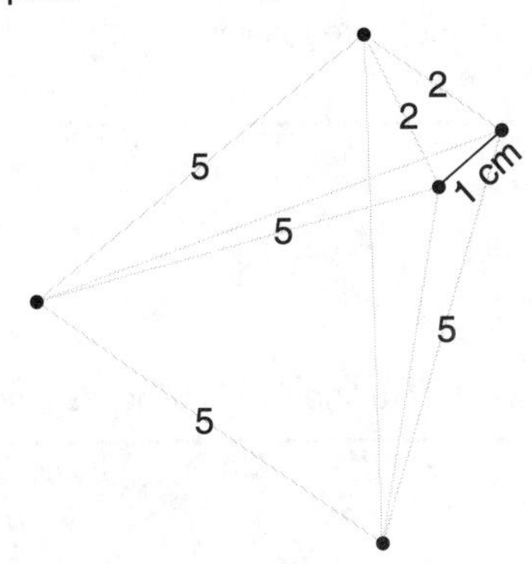

The only way to use these lines to complete the path in the order given is shown below.

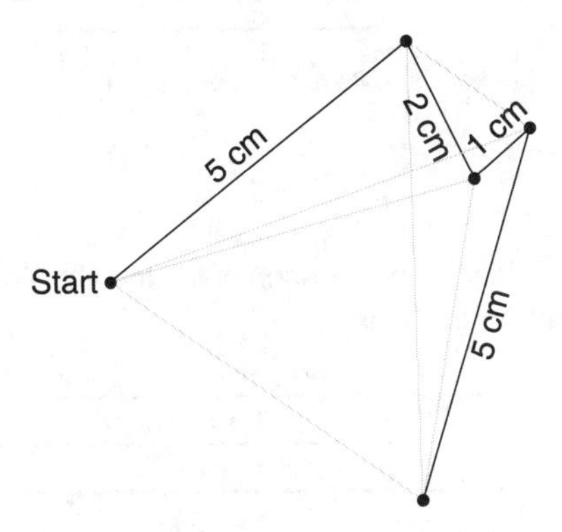

41. Since there is only one line connecting two dots that are 2 cm apart, it must be part of the path.

Since there are only two lines connecting two dots that are 7 cm apart, both must be part of the path.

We trace these lines as shown, then connect the only 4 cm line that completes the path.

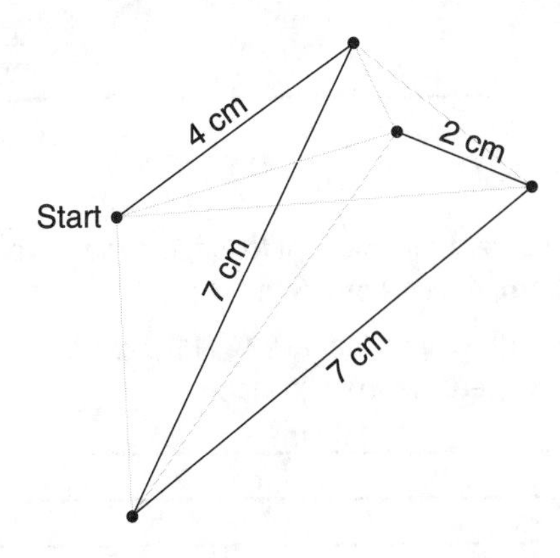

42. Since there are only three lines connecting two dots that are 5 cm apart, all of these lines must be part of the path. We trace these lines as shown.

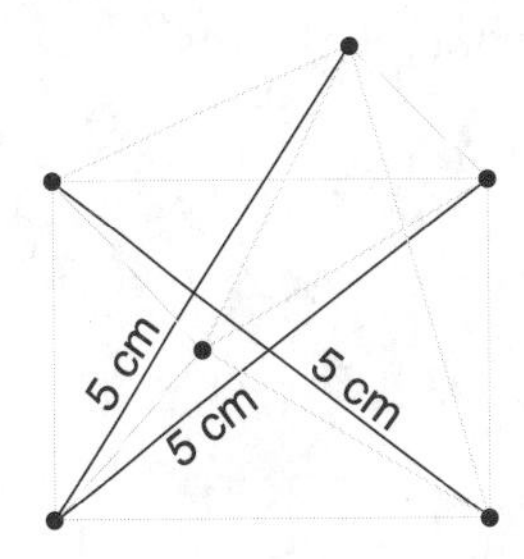

The path begins with two 5 cm lines. We trace the only 4 cm line connecting these two 5 cm lines to the other 5 cm line.

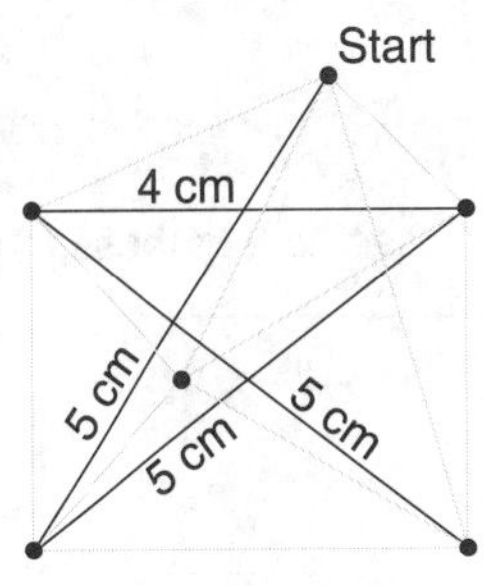

Finally, we complete the path by connecting the only remaining dot with a 3 cm line.

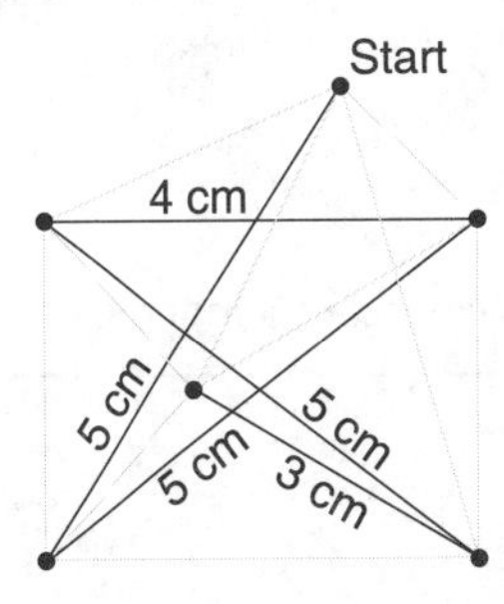

43. Since there are only two lines connecting two dots that are 3 cm apart, both must be part of the path. We trace these lines as shown. These are the end of our path.

Then, we label the lines connecting dots that are 5 or 6 cm apart.

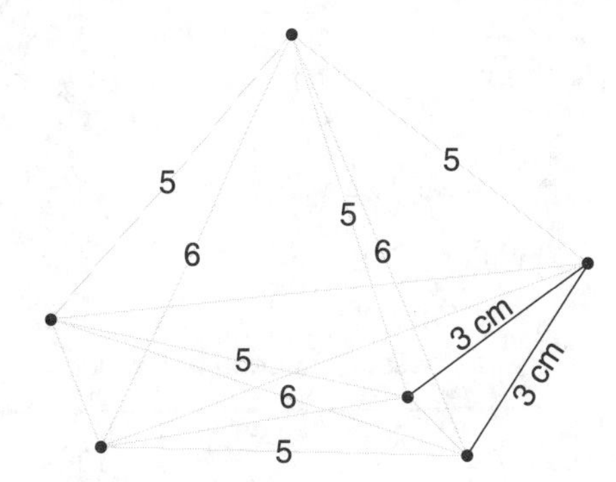

Working backwards from the two 3 cm lines, we look for a path connecting the three remaining dots with lines that are 6 cm, 5 cm, and 6 cm. There is only one way to do this, as shown.

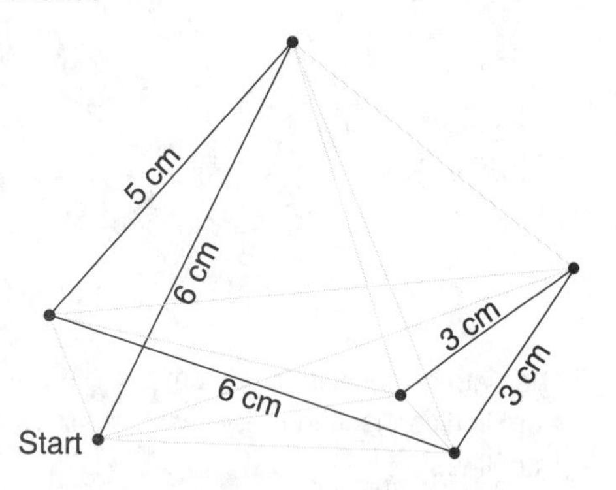

MEASUREMENT

Nearest Measure 20–21

44. The line is closest to **3 inches** long.

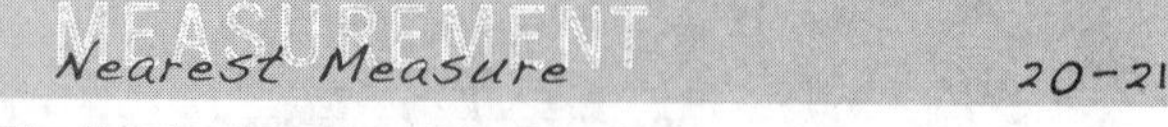

45. The line is closest to **2 inches** long.

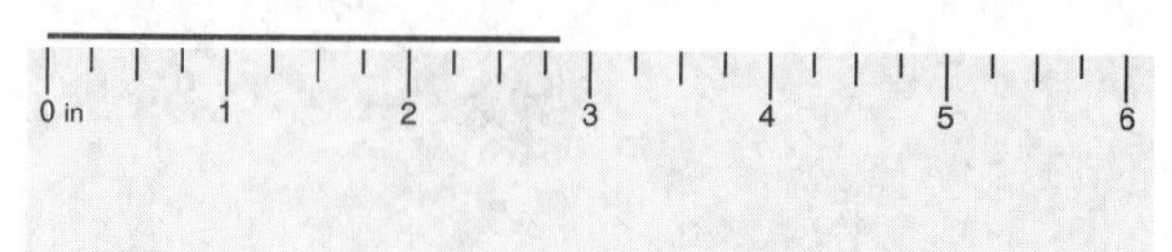

46. The line is closest to **5 inches** long.

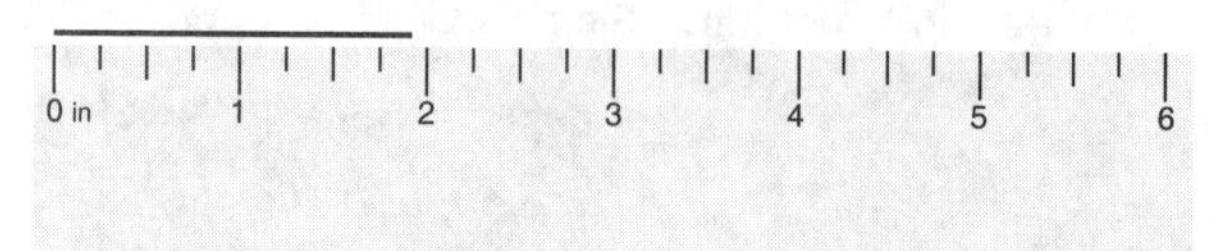

47. The line is closest to **7 centimeters** long.

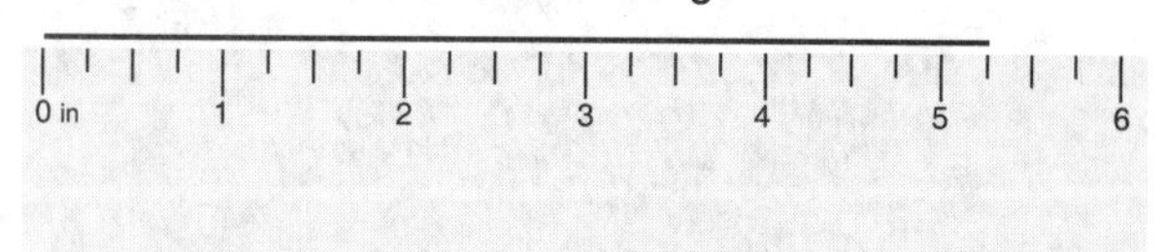

48. The line is closest to **5 centimeters** long.

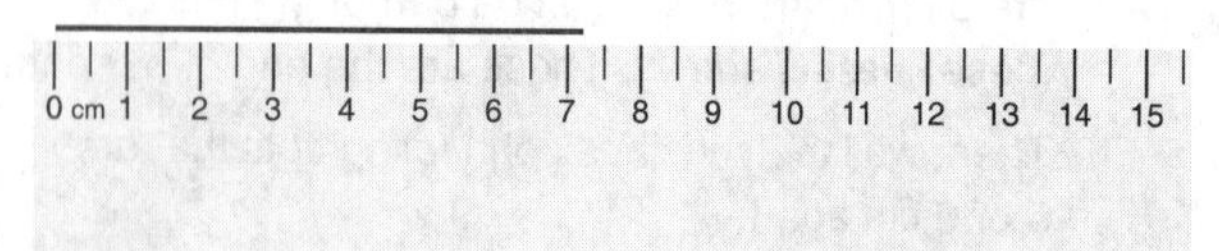

49. The line is closest to **13 centimeters** long.

50. First, we draw a line that is 15 centimeters long.

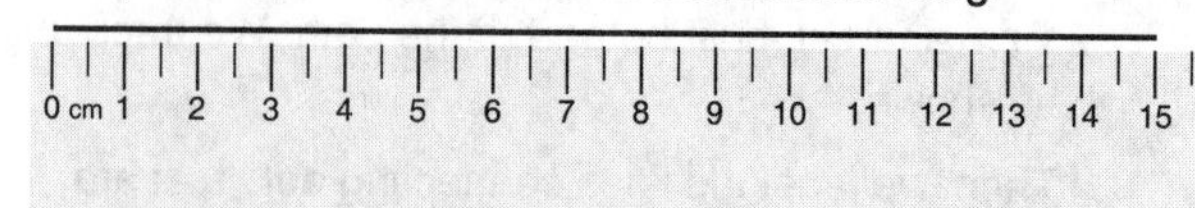

Then, we measure the line in inches. The line is about **6** inches long.

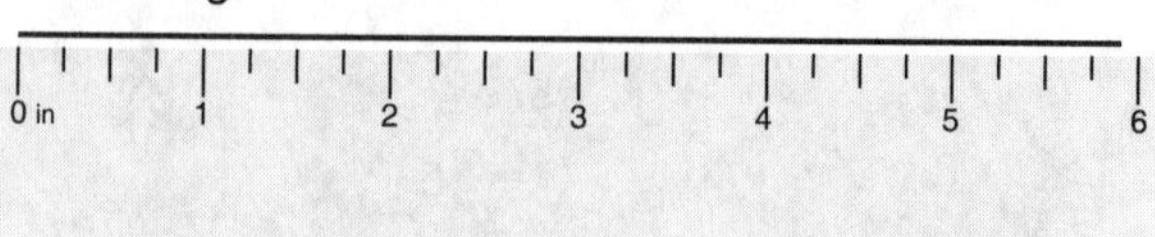

51. First, we draw a line that is 5 inches long.

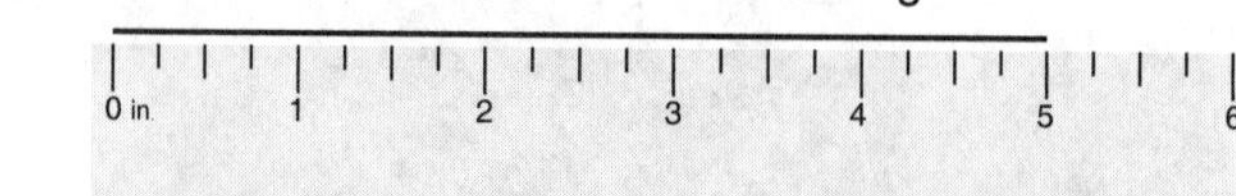

Then, we measure the line in centimeters. The line is about **13** centimeters long.

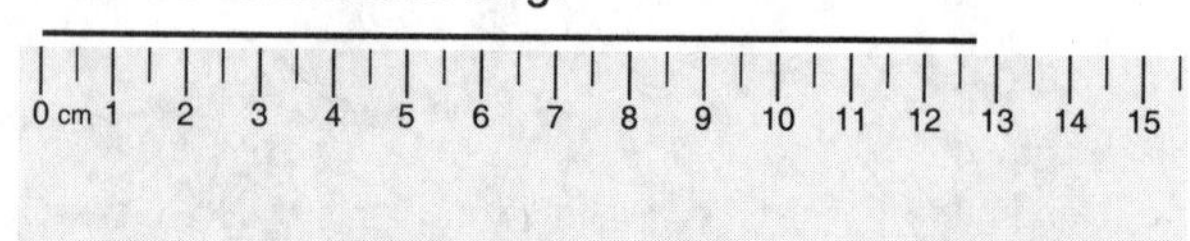

52. We draw a line for each of the lengths, lining up the left ends of the lines.

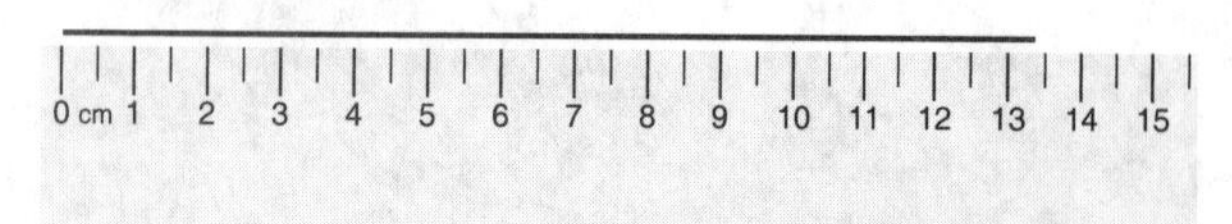

In order from shortest to longest, we have:
9 cm, 4 in, 12 cm, 5 in.

— or —

We draw one line and make marks at 9 cm, 12 cm, 4 in, and 5 in.

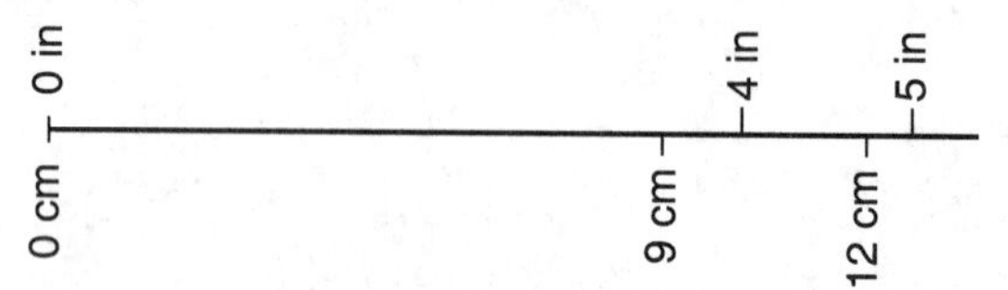

In order from shortest to longest, we have:
9 cm, 4 in, 12 cm, 5 in.

53. We draw lines that are 10, 11, 12, 13, 14, and 15 cm long and measure them in inches.

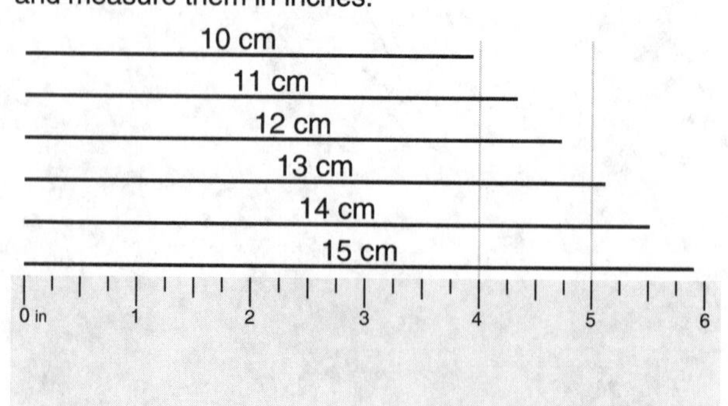

Only the 11 cm and 12 cm lines are between 4 and 5 inches long. So, **2** of Alex's lines are between 4 and 5 inches long.

— or —

We draw one line and make marks at 10, 11, 12, 13, 14, and 15 cm. Then, we make marks at 4 in and 5 in and compare.

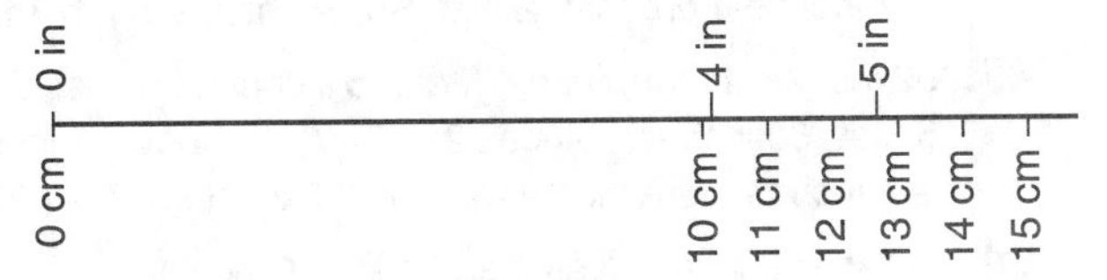

Only the 11 cm and 12 cm marks are between the marks for 4 in and 5 in.

So, **2** of Alex's lines are between 4 and 5 inches long.

54. We can add ten lengths of 4 in to get a length of 40 in.

Since 4 in is about 10 cm, this gives us a total of about 10+10+10+10+10+10+10+10+10+10 = 100 cm.

16 centimeters 50 centimeters 80 centimeters 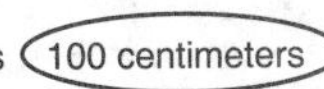

MEASUREMENT Other Units of Length 22-23

55. Most refrigerators are about **6 feet** tall.

56. Most home ceilings are about **9 feet** high.

57. A community pool is usually about **25 yards** long.

58. A square baking dish is usually about **9 inches** wide.

59. Towns can be any number of miles apart. Since none of the other answers make sense, we choose **15 miles**.

We connect our answers for Problems 55-59 as shown.

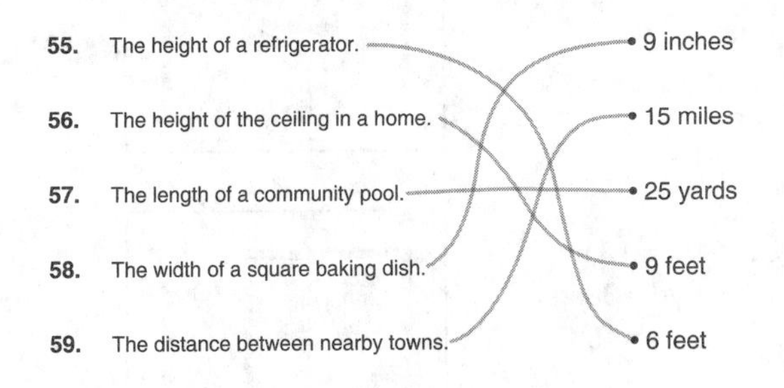

60. Most clothes hangers are about **45 centimeters** long.

61. Most beds are about **2 meters** long.

62. Pens are usually about **15 centimeters** long.

63. An airport runway is much longer than 8 meters long. The only answer that makes sense is **2 kilometers**.

64. A flagpole is usually taller than 2 meters, but much shorter than 2 kilometers. The best answer choice is the only one that is left, **8 meters**.

We connect our answers for Problems 60-64 as shown.

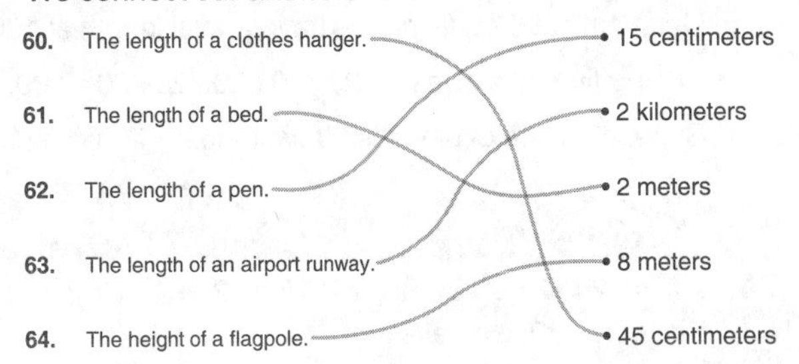

MEASUREMENT Changing Units 24-25

65. Since 1 yard equals 3 feet, 4 yards equals 3+3+3+3 = **12** feet.

66. Since 1 foot is 12 inches, we count how many times we add 12 inches to get 60 inches. Adding five 12's gives us 12+12+12+12+12 = 60.

So, **5** feet equals 12+12+12+12+12 = 60 inches.

67. To walk the whole way around the square, Toby walks 6+6+6+6 = 24 feet.

Since 1 yard is 3 feet, we count how many times we add 3 feet to get 24 feet.

Adding eight 3's gives us 3+3+3+3+3+3+3+3 = 24.

So, **8** yards equals 3+3+3+3+3+3+3+3 = 24 feet.

— or —

Each side of the square is 6 feet long. Since 1 yard equals 3 feet, 2 yards equals 3+3 = 6 feet.
So, each side of the square is 2 yards long.

Walking all four sides of the square, Toby walks a total of 2+2+2+2 = **8** yards.

68. Since 1 meter equals 100 cm,
3 meters equals 100+100+100 = **300** cm.

69. First, we change yards to feet.

Since 1 yard equals 3 feet, 2 yards equals 3+3 = 6 feet.

Then, we change feet to inches.

Since 1 foot equals 12 inches, 6 feet equals 12+12+12+12+12+12 = **72** inches.

— or —

1 yard equals 3 feet, and 1 foot equals 12 inches.

So, 1 yard equals 12+12+12 = 36 inches.

This means 2 yards is 36+36 = **72** inches.

70. 100 centimeters equals 1 meter.
Since 600 = 100+100+100+100+100+100,
600 centimeters equals **6** meters.

71. Tog's boat is 6 yards long. Since 1 yard is 3 feet, Tog's boat is 3+3+3+3+3+3 = 18 feet long.

So, Dar's 20-foot boat is 20 − 18 = **2** feet longer than Tog's boat.

72. Since 1 foot is 12 inches, 4 feet is $12+12+12+12=48$ inches. So, Herb's tail is $55-48=$ **7** inches longer than Gerb's.

73. 1 meter equals 100 centimeters. We look for a length of string so that 5 equal pieces have a total length of 100 cm.

Adding five 20's gives us $20+20+20+20+20=100$.

So, each piece of string is **20** cm long.

MEASUREMENT
Mixed Measures 26–27

74. 2 ft is $12+12=24$ in. So, 2 ft 3 in is $24+3=27$ in. Since $27<30$, we have

2 ft 3 in (<) 30 in.

75. 3 ft is $12+12+12=36$ in. So, 3 ft 6 in is $36+6=42$ in. Since $40<42$, we have

40 in (<) 3 ft 6 in.

76. 1 ft 3 in is $12+3=15$ in. Since $15=15$, we have

15 in (=) 1 ft 3 in.

77. 10 ft is $12+12+12+12+12+12+12+12+12+12=120$ in. So, 10 ft 1 in is $120+1=121$ in.

Since $121>101$, we have

10 ft 1 in (>) 101 in.

— or —

To change 10 ft 1 in to inches, we add ten 12's plus 1 inch.

101 inches is ten 10's plus 1 inch.

Since ten 12's is more than ten 10's, we have

10 ft 1 in (>) 101 in.

78. 7 ft 8 in is 7 feet plus 8 inches.
8 feet is 7 feet plus 12 inches.

Since 8 feet is more than 7 ft 8 in, we have

7 ft 8 in (<) 8 ft 7 in.

79. 2 yards is $3+3=6$ feet, which is 5 feet plus 12 inches. Since 5 ft 12 in is more than 5 ft 11 in, we have

2 yards (>) 5 ft 11 in.

80. To get from 3 feet 3 inches to 4 feet 4 inches, we add 1 foot and 1 inch.

Since 1 foot is 12 inches, 1 foot and 1 inch is $12+1=13$ inches.

So, 4 ft 4 in is **13** inches longer than 3 ft 3 in.

81. 3 feet is $12+12+12=36$ inches.
So, 3 feet 6 inches is $36+6=42$ inches.

6 feet is $12+12+12+12+12+12=72$ inches.
So, 6 feet 3 inches is $72+3=75$ inches.

Since $75-42=33$, we know 75 inches is **33** inches more than 42 inches.

— or —

To get from 3 ft 6 in to 6 ft 3 in, we can add 3 feet then take away 3 inches.

Adding 3 feet is the same as adding $12+12+12=36$ inches. So, adding 3 feet then taking away 3 inches is the same as adding $36-3=$ **33** inches.

82. 31 inches is more than 2 feet (24 inches), but less than 3 feet (36 inches). Since $31-24=7$, we know 31 inches is 7 inches more than 2 feet. So, 31 inches is **2 ft 7 in.**

83. 52 inches is more than 4 feet (48 inches), but less than 5 feet (60 inches). Since $52-48=4$, we know 52 inches is 4 inches more than 4 feet. So, 52 inches is **4 ft 4 in.**

84. Julian is 3 inches less than 5 feet (60 inches) tall. So, Julian is $60-3=57$ inches tall. Navin is 5 inches taller than Julian, which is $57+5=62$ inches tall.

Since 5 feet is 60 inches, Navin is $62-60=2$ inches taller than 5 feet, which is **5 ft 2 in.**

— or —

If Navin were 3 inches taller than Julian, he would be exactly 5 feet tall. Since he is 5 inches taller than Julian, he is $5-3=2$ inches taller than 5 feet tall, or **5 ft 2 in.**

85. In a mixed measure, the number of inches should always be less than 12.

14 feet 14 inches is 14 feet + 12 inches + 2 inches, which is 14 feet + 1 foot + 2 inches, or **15 ft 2 in.**

86. 5 ft = 60 in
6 ft = 2 yd
12 ft = 4 yd

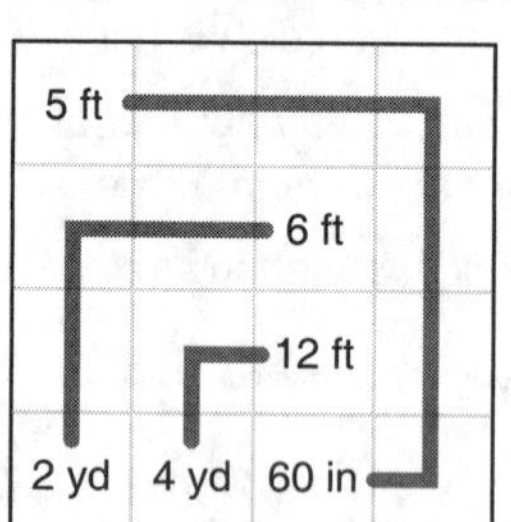

87. 1 ft = 12 in
1 yd = 36 in
5 yd = 15 ft

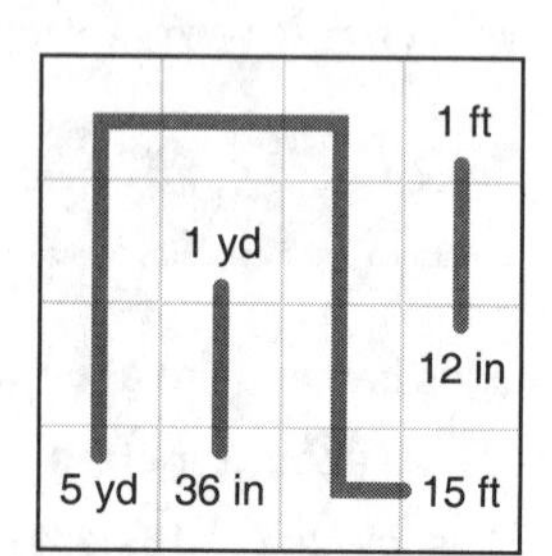

88. 1 yd = 3 ft
24 in = 2 ft
1 ft = 12 in

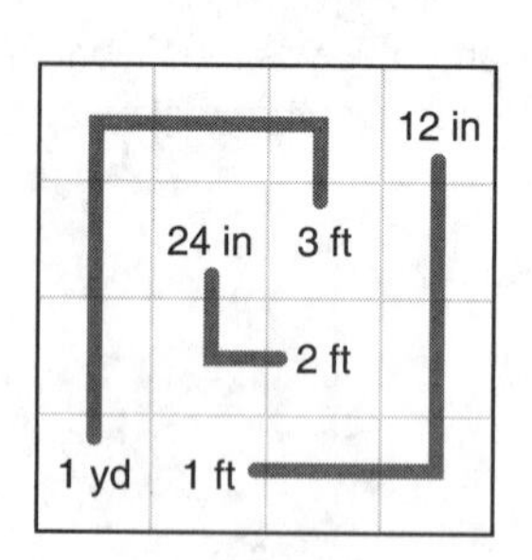

89. 9 ft = 3 yd
4 yd = 12 ft
48 in = 4 ft

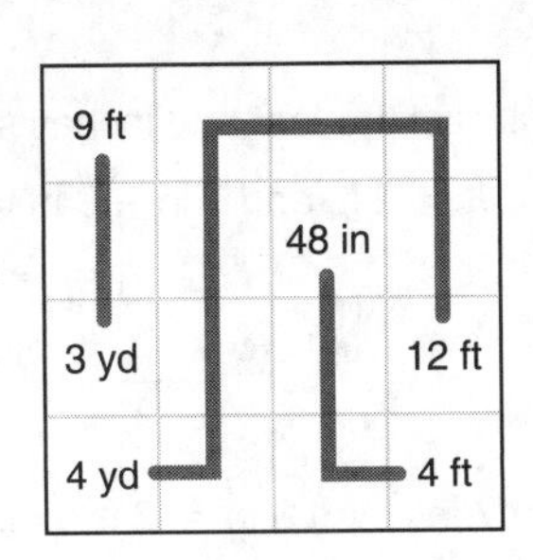

90. 36 in = 3 ft
6 ft = 2 yd
6 yd = 18 ft

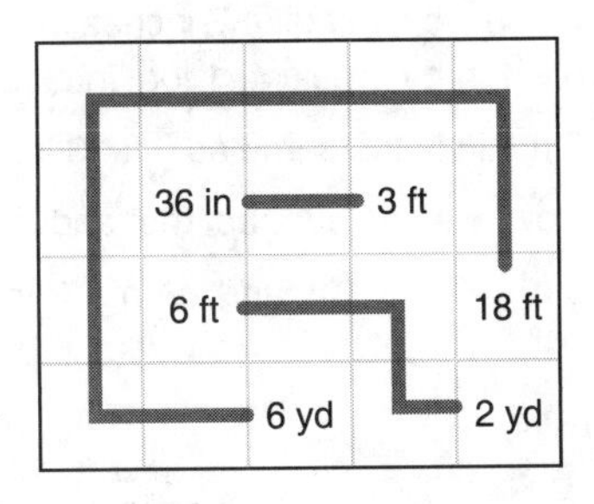

91. 9 ft = 3 yd
2 ft = 24 in
5 yd = 15 ft

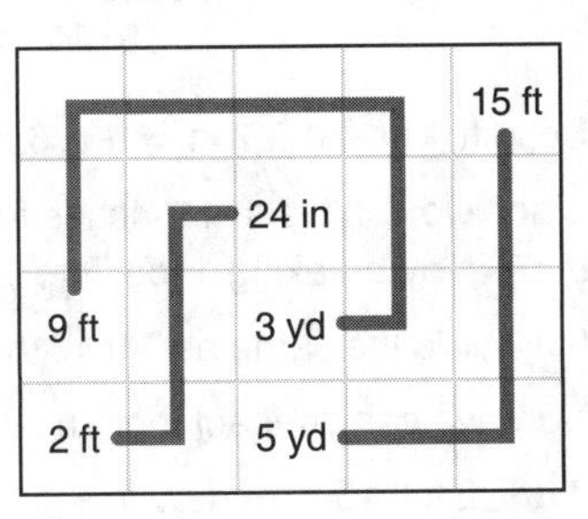

92. 3 ft = 1 yd
60 in = 5 ft
4 ft = 48 in

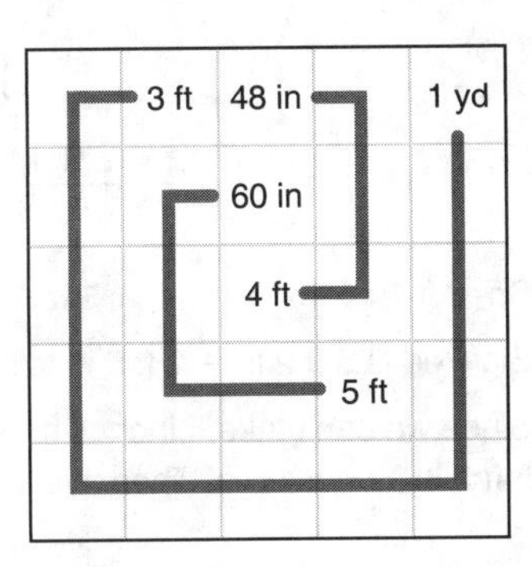

93. 12 in = 1 ft
4 yd = 12 ft
1 yd = 36 in

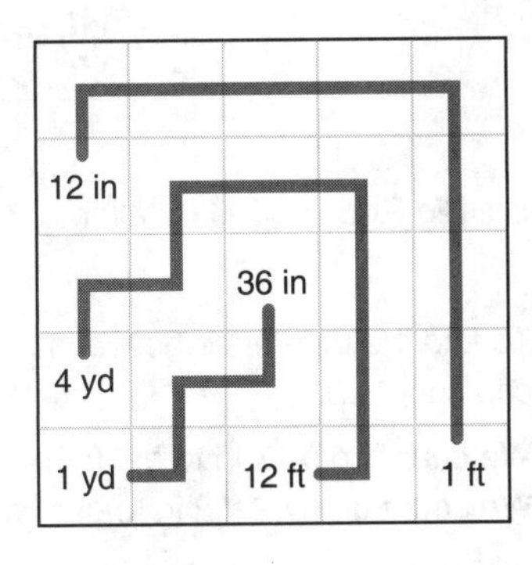

94. 21 ft = 7 yd
72 in = 2 yd
10 ft = 120 in

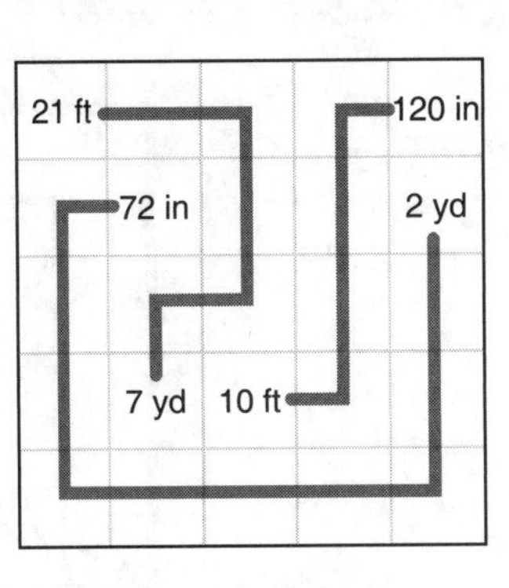

95. 36 in = 1 yd
10 yd = 30 ft
4 yd = 12 ft

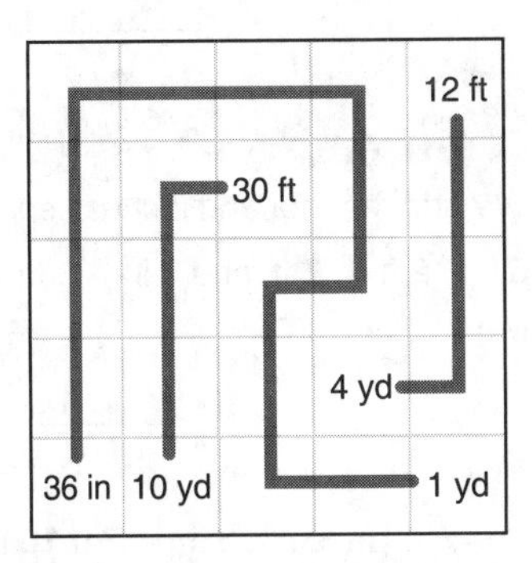

96. 5 yd = 15 ft
6 ft = 2 yd
36 in = 3 ft
48 in = 4 ft

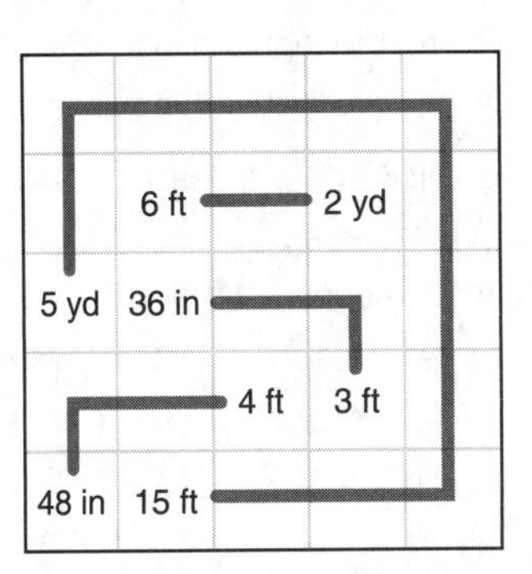

97. 9 ft = 3 yd
24 in = 2 ft
6 yd = 18 ft
60 in = 5 ft

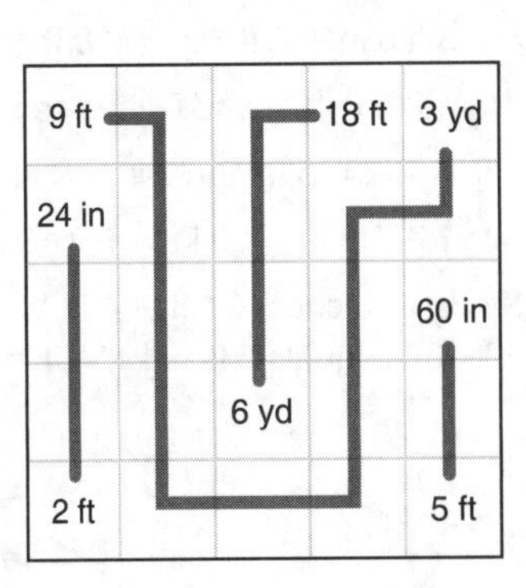

98. 6 ft = 2 yd
8 yd = 24 ft
48 in = 4 ft
360 in = 10 yd

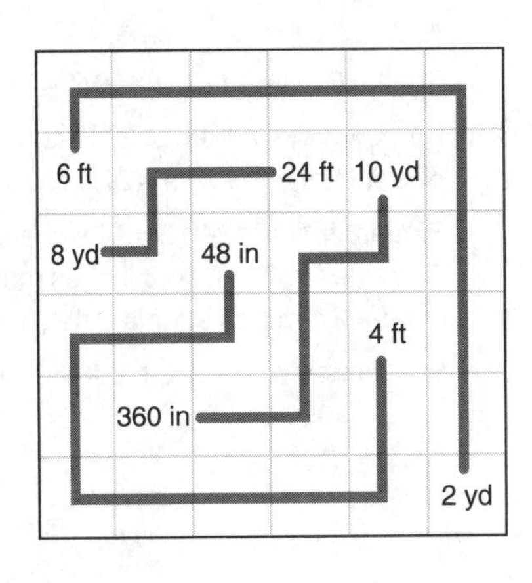

99. 36 in = 3 ft
240 in = 20 ft
9 yd = 27 ft
72 in = 2 yd

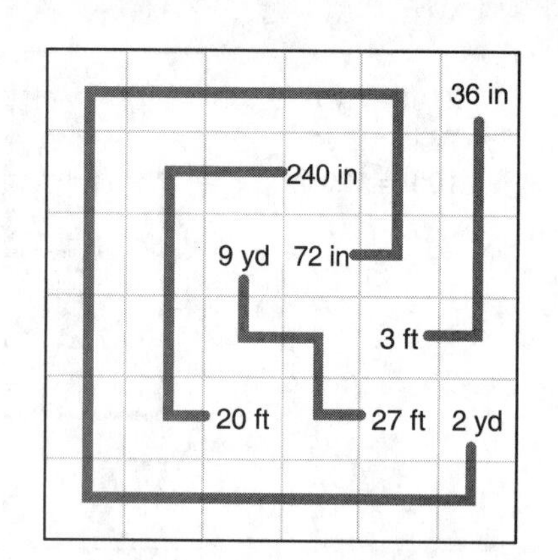

100. We add the feet and inches separately.

5 ft + 2 ft = 7 ft, and 7 in + 3 in = 10 in.

```
   5 ft  7 in
+  2 ft  3 in
-------------
   7 ft 10 in
```

So, 5 ft 7 in + 2 ft 3 in = **7 ft 10 in.**

101. We add the feet and inches separately.

3 ft + 6 ft = 9 ft, and 8 in + 5 in = 13 in.

So, 3 ft 8 in + 6 ft 5 in = 9 ft 13 in.

Since 13 inches is 1 ft + 1 in, we know that 9 ft 13 in equals 9 ft + 1 ft + 1 in, which is 10 ft 1 in.

```
   3 ft  8 in
+  6 ft  5 in
-------------
   9 ft 13 in  (crossed out)
  10 ft  1 in
```

So, 3 ft 8 in + 6 ft 5 in = **10 ft 1 in.**

102. We add the feet and inches separately.

4 ft + 1 ft = 5 ft, and 7 in + 9 in = 16 in.

So, 4 ft 7 in + 1 ft 9 in = 5 ft 16 in.

Since 16 inches is 1 ft + 4 in, we know that 5 ft 16 in equals 5 ft + 1 ft + 4 in, which is 6 ft 4 in.

```
   4 ft  7 in
+  1 ft  9 in
-------------
   5 ft 16 in  (crossed out)
   6 ft  4 in
```

So, 4 ft 7 in + 1 ft 9 in = **6 ft 4 in.**

103. Barney walks a total of 3+3+3 = 9 feet, plus a total of 9+9+9 = 27 inches.

Since 2 feet is 12+12 = 24 inches, 27 inches is 27−24 = 3 inches more than 2 feet, which means 27 inches equals 2 ft + 3 in.

So, Barney walks 9 ft + 2 ft + 3 in = **11 ft 3 in.**

104. We subtract the feet and inches separately.

3 ft − 1 ft = 2 ft, and 7 in − 3 in = 4 in.

```
   3 ft  7 in
-  1 ft  3 in
-------------
   2 ft  4 in
```

So, 3 ft 7 in − 1 ft 3 in = **2 ft 4 in.**

105. Since we can't take 7 inches from 1 inch, we rewrite 6 ft 1 in by breaking 1 foot into 12 inches.

6 ft 1 in is the same as 5 ft 13 in.

Now, we can subtract feet and inches separately.

5 ft − 4 ft = 1 ft, and 13 in − 7 in = 6 in.

```
   5 ft 13 in
   6 ft  1 in  (crossed out)
-  4 ft  7 in
-------------
   1 ft  6 in
```

So, 6 ft 1 in − 4 ft 7 in = **1 ft 6 in.**

106. Since we can't take 11 inches from 7 inches, we rewrite 8 ft 7 in by breaking 1 foot into 12 inches.

8 ft 7 in is the same as 7 ft 19 in.

Now, we can subtract feet and inches separately.

7 ft − 2 ft = 5 ft, and 19 in − 11 in = 8 in.

```
   7 ft 19 in
   8 ft  7 in  (crossed out)
-  2 ft 11 in
-------------
   5 ft  8 in
```

So, 8 ft 7 in − 2 ft 11 in = **5 ft 8 in.**

107. Big Red is 5 ft 2 in − 2 ft 5 in taller than Nellie.

Since we can't take 5 inches from 2 inches, we rewrite 5 ft 2 in as 4 ft 14 in. Then, we subtract.

```
   4 ft 14 in
   5 ft  2 in  (crossed out)
-  2 ft  5 in
-------------
   2 ft  9 in
```

So, Big Red is **2 ft 9 in** taller.

— or —

To find the difference between 5 ft 2 in and 2 ft 5 in, we can count up.

We count up by 7 in to get from 2 ft 5 in to 3 ft.
We count up by 2 ft 2 in to get from 3 ft to 5 ft 2 in.

All together, we count up by 7 in + 2 ft 2 in = 2 ft 9 in.
So, Big Red is **2 ft 9 in** taller.

108. The height of the stack is 8+8+8+8+8+8+8 = 56 inches.

56 inches is more than 4 feet (48 inches), but less than 5 feet (60 inches). Since 56−48 = 8, we know 56 inches is 8 inches more than 4 feet. So, 56 inches is **4 ft 8 in.**

— or —

Three blocks are $8+8+8=24$ inches tall, which is 2 feet.

So, six blocks are $2+2=4$ feet tall.

Seven blocks are 4 feet + 8 inches tall, which is **4 ft 8 in**.

109. The distance Ernie has left to travel is 7 ft − 4 ft 5 in.

We write 7 ft as 6 ft 12 in, then subtract.

$$\begin{array}{r} 6\text{ ft } 12\text{ in} \\ -\ 4\text{ ft } \ 5\text{ in} \\ \hline 2\text{ ft } \ 7\text{ in} \end{array}$$

So, Ernie has **2 ft 7 in** left to travel.

— or —

To find the difference between 7 ft and 4 ft 5 in, we can count up.

We count up by 7 in to get from 4 ft 5 in to 5 ft.
We count up by 2 ft to get from 5 ft to 7 ft.

All together, we count up by 7 in + 2 ft = 2 ft 7 in.
So, Ernie has **2 ft 7 in** left to travel.

110. We can write 20 inches as 12 in + 8 in = 1 ft 8 in.

We can write 20 feet as 19 ft + 12 in = 19 ft 12 in.

So, cutting 20 in from 20 ft leaves 19 ft 12 in − 1 ft 8 in.

$$\begin{array}{r} 19\text{ ft } 12\text{ in} \\ -\ 1\text{ ft } \ 8\text{ in} \\ \hline 18\text{ ft } \ 4\text{ in} \end{array}$$

So, there are **18 ft 4 in** of rope left.

— or —

20 inches is 1 ft 8 in. To find the difference between 20 ft and 1 ft 8 in, we can count up.

We count up by 4 in to get from 1 ft 8 in to 2 ft.
We count up by 18 ft to get from 2 ft to 20 ft.

All together, we count up by 4 in + 18 ft = 18 ft 4 in.
So, there are **18 ft 4 in** of rope left.

111. We can compute the feet and inches separately.

5 ft + 4 ft − 6 ft = 3 ft.

8 in + 7 in − 5 in = 10 in.

So, 5 ft 8 in + 4 ft 7 in − 6 ft 5 in = **3 ft 10 in**.

MEASUREMENT Challenge Problems 34–35

112. 11 feet is the same as 10 feet and 12 inches.

So, 11 feet is 10 feet and 1 inch longer than 11 inches.

To write 10 ft 1 in as inches, we add ten 12's to get 120, plus 1 inch to get **121** inches.

113. 4 yards is $3+3+3+3=12$ feet. Subtracting 5 feet leaves $12-5=7$ feet.

Then, 7 feet is $12+12+12+12+12+12+12=$ **84** inches.

114. The 8 sides of the stop sign have a total length of 2 yards, which is $3+3=6$ feet.

6 feet is $12+12+12+12+12+12=72$ inches. We look for a side length so that eight equal sides have a total length of 72 inches.

Adding eight 9's gives us $9+9+9+9+9+9+9+9=72$. So, eight 9-inch sides have a total length of 72 inches.

Each side is **9** inches long.

— or —

The 8 sides of the stop sign have a total length of 2 yards. So, the length of 4 sides of the stop sign is half as much, or 1 yard.

1 yard is $12+12+12=36$ inches. We look for a side length so that four equal sides have a total length of 36 inches.

Adding four 9's gives us $9+9+9+9=36$. So, four 9-inch sides have a total length of 36 inches.

Each side is **9** inches long.

115. We show the length of all six measurements on the page below.

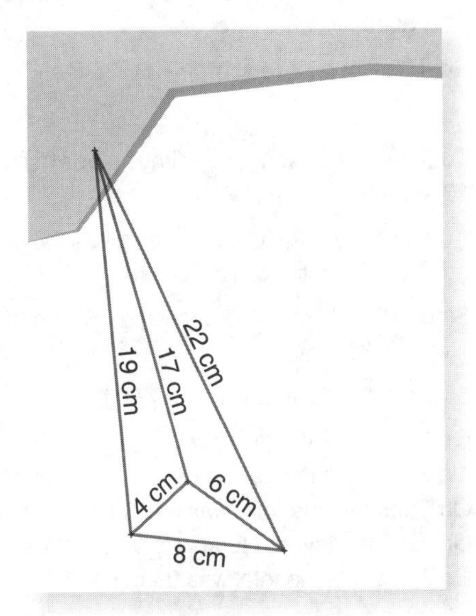

The sum of these six lengths is $22+17+19+4+6+8$ cm.

We pair numbers that are easy to add and compute: $(22+8)+(4+6)+17+19=$ **76** cm.

116. The short line is between 5 and 6 cm long.

The long line is between 12 and 13 cm long.

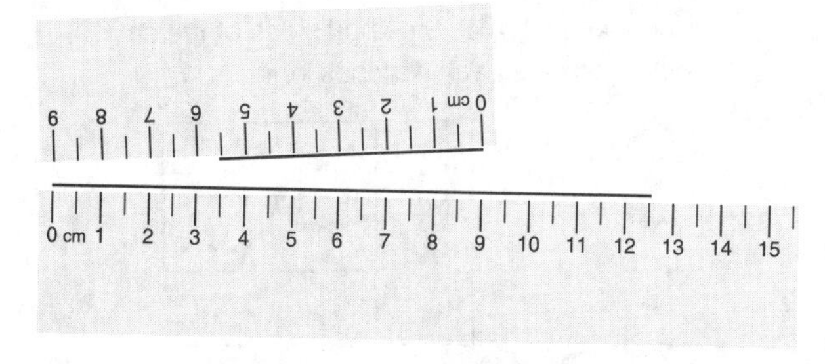

To find out how much longer the long line is than the short line, we place a mark on the long line at the length of the short line.

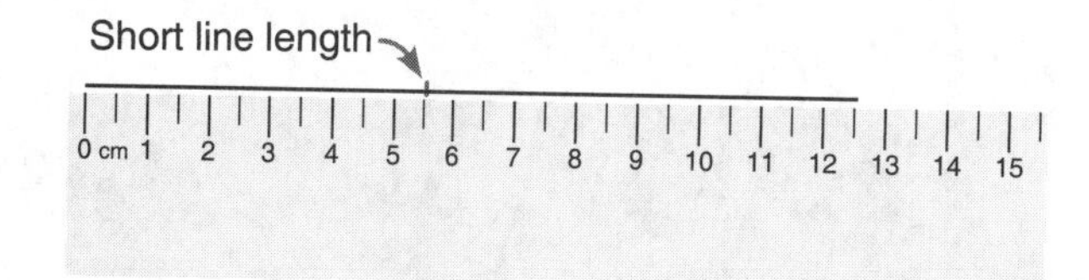

Then, we measure the length from this mark to the end of the long line.

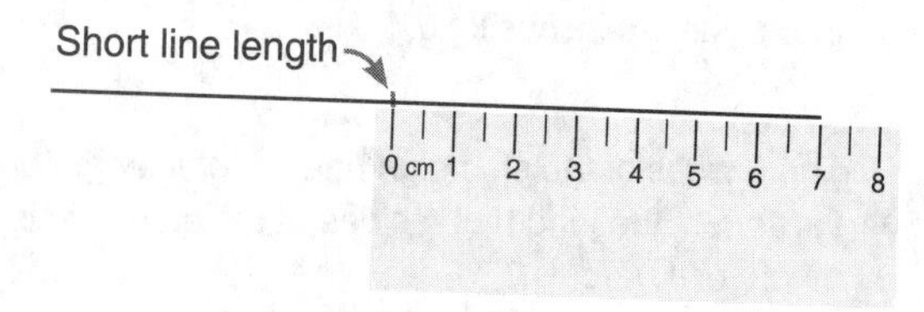

The long line is **7** cm longer than the short line.

117. 3 feet is $12+12+12=36$ inches.

So, 35 inches is 1 inch less than 3 feet.

With each hop, Kirby travels 3 feet, minus 1 inch.

So, in 12 hops, Kirby will travel $3+3+3+3+3+3+3+3+3+3+3+3=36$ feet, minus 12 inches, which is 1 foot.

So, Kirby hops a total of $36-1=$ **35** feet.

— *or* —

To find the total length of Kirby's hops in inches, we add 12 lengths of 35 inches.

Adding 12 lengths of 35 inches gives the same result as adding 35 lengths of 12 inches.

Since 12 inches is 1 foot, adding 35 lengths of 12 inches gives us **35** feet.

118. 5 feet is $12+12+12+12+12=60$ inches.

On Sunday, the total snowfall was 9 inches.
On Monday, the total was $9+10=19$ inches.
On Tuesday, the total was $9+10+11=30$ inches.
On Wednesday, the total was $9+10+11+12=42$ inches.
On Thursday, the total was $9+10+11+12+13=55$ inches.

On Friday, it snowed 14 more inches, bringing the total to $55+14=69$ inches, which is more than 5 feet.

So, the snowfall passed 5 feet on **Friday**.

119. First, we find the length of a short side of a small rectangle. Together, three short sides of a small rectangle equal one long side of a small rectangle. Each long side is 1 foot, or 12 inches long.

Since $4+4+4=12$, the short sides of the small rectangles are each 4 inches long.

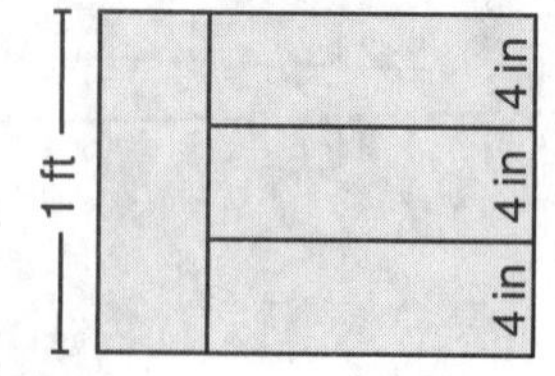

Since all of the rectangles are the same, we can label any of the short sides 4 inches and any of the long sides 12 inches. We label two sides below to find the length of the long side of the big rectangle.

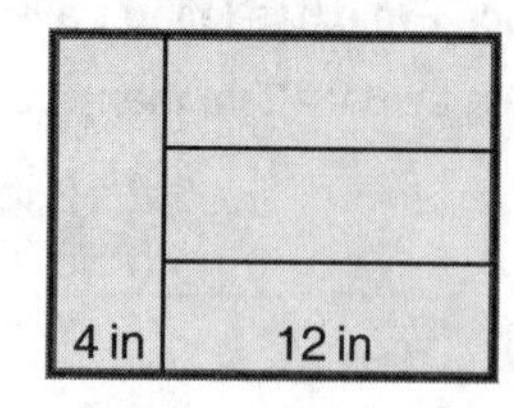

So, the long side of the big rectangle is $4+12=$ **16** inches.

STRATEGIES Review 37–39

1. $9+8-3 = 17-3 =$ **14.**

2. $15-6+4 = 9+4 =$ **13.**

3. $11-8+4-5 = 3+4-5 = 7-5 =$ **2.**

4. $8-1+2+6 = 7+2+6 = 9+6 =$ **15.**

5. $47+23-14 = 70-14 =$ **56.**

6. $100-68+32 = 32+32 =$ **64.**

7. $87-37+25-11 = 50+25-11 = 75-11 =$ **64.**

8. $24+36+18-31 = 60+18-31 = 78-31 =$ **47.**

9. $18+12-7-8+2 = 30-7-8+2 = 23-8+2 = 15+2 =$ **17.**

10. $350-40-200-40+110 = 310-200-40+110 = 110-40+110 = 70+110 =$ **180.**

11. $32+49+51 = 32+100 =$ **132.**

12. $17+44+23 = 40+44 =$ **84.**

13. $25+39+25 = 50+39 =$ **89.**

14. $13+18+17+12 = 30+30 =$ **60.**

15. $111+89+16+24 = 200+40 =$ **240.**

16. $35+35+65+65 = 100+100 =$ **200.**

17. $126+32+74+118 = 200+150 =$ **350.**

18. $12+245+99+118+105 = 130+350+99 = 480+99 =$ **579.**

19. We can make five pairs of numbers that sum to 100.

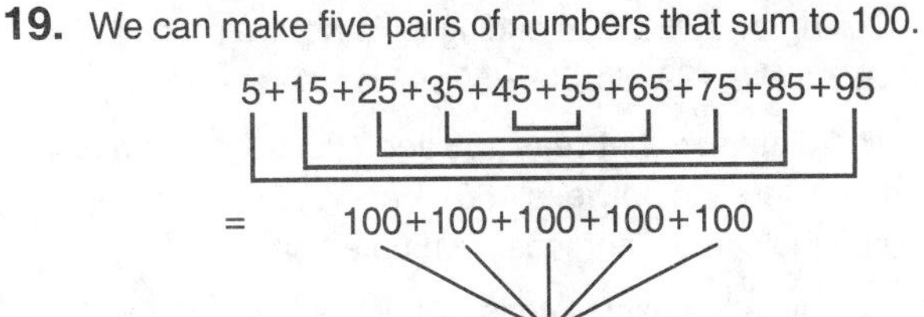

$5+15+25+35+45+55+65+75+85+95$

$= 100+100+100+100+100$

$=$ **500.**

20. We change the order of the numbers being subtracted to make our computation easier:

$86-9-56 = 86-56-9$
$= 30-9$
$=$ **21.**

21. Taking away 25 then taking away 25 more is the same as taking away 50 all at once.

So, $579-25-25 = 579-50 =$ **529.**

22. We change the order of the numbers being subtracted to make our computation easier:

$276-39-26 = 276-26-39$
$= 250-39$
$=$ **211.**

23. We change the order of the numbers being subtracted to make our computation easier:

$353-37-63 = 353-63-37$
$= 290-37$
$=$ **253.**

— *or* —

Taking away 37 then taking away 63 is the same as taking away 100 all at once.

So, $353-37-63 = 353-100 =$ **253.**

24. We consider each choice.

Taking away 28 then adding 12 is *not* the same as taking away 12 then taking away 28.
So, $88-28+12$ is not equal to $88-12-28$. ✘

Adding 12 then taking away 28 is *not* the same as taking away 12 then taking away 28.
So, $88+12-28$ is not equal to $88-12-28$. ✘

Taking away 28 then taking away 12 is the same as taking away 12 then taking away 28.
So, $88-28-12 = 88-12-28$. ✓

Adding 40 is *not* the same as taking away 12 then taking away 28. So, $88+40$ is not equal to $88-12-28$. ✘

Taking away 40 is the same as taking away 12 then taking away 28.
So, $88-40 = 88-12-28$. ✓

We circle the expressions that are equal to $88-12-28$.

$88-28+12$ $88+12-28$ ($88-28-12$ circled)

$88+40$ ($88-40$ circled)

25. We consider each choice.

Taking away 100 then adding 1 is the same as taking away 99. So, $155-100+1=155-99$. ✓

Taking away 55 then taking away 44 is the same as taking away 99 all at once. So, $155-55-44=155-99$. ✓

Taking away 55 then adding 44 is *not* the same as taking away 99. So, $155-55+44$ is not equal to $155-99$. ✗

The difference between 156 and 100 is the same as the difference between $156-1$ and $100-1$. So, $156-100$ has the same difference as $155-99$. ✓

We circle the expressions that are equal to $155-99$.

(155−100+1) (155−55−44)
155−55+44 (156−100)

26. After Elmer adds 180 squares then removes 44 squares, the number of squares remaining is $144+180-44=324-44=\mathbf{280}$.

— or —

The quilt starts with 144 squares. Adding 180 squares then removing 44 squares leaves the same number of squares as removing 44 squares then adding 180 squares.

Since subtracting 44 first is easier, we compute $144-44+180=100+180=\mathbf{280}$ squares.

27. The pile starts with 125 bananas. Eating 48 bananas then adding 25 bananas leaves the same number of bananas as adding 25 bananas then eating 48 bananas.

Since adding 25 first is easier, we compute $125+25-48=150-48=\mathbf{102}$ bananas.

28. Buster starts with 141 flamingoats. Selling 23 flamingoats, then losing 31 flamingoats, then finding 20 flamingoats leaves the same number of flamingoats as losing 31 flamingoats, then finding 20 flamingoats, then selling 23 flamingoats.

Since subtracting 31 first is easier, we compute $141-31+20-23=110+20-23=130-23=\mathbf{107}$ flamingoats.

29. Adding 39 then subtracting 32 is the same as subtracting 32 then adding 39.

So, $32+39-32=32-32+39$.

32+32−39 (32−32+39) 32−32−39 32+32+39

30. Subtracting 16 then adding 45 is the same as adding 45 then subtracting 16.

So, $55-16+45=55+45-16$.

55+45+16 55−45−16 (55+45−16) 55−45+16

31. Subtracting 18 then adding 44 is the same as adding 44 then subtracting 18.

So, $44-18+44=44+44-18$.

(44+44−18) 44−44−18 44+44+18 44−44+18

32. We use our answers from the three previous problems to evaluate each expression as shown.

$$\begin{aligned}32+39-32&=32-32+39\\&=0+39\\&=\mathbf{39}.\end{aligned}$$

$$\begin{aligned}55-16+45&=55+45-16\\&=100-16\\&=\mathbf{84}.\end{aligned}$$

$$\begin{aligned}44-18+44&=44+44-18\\&=88-18\\&=\mathbf{70}.\end{aligned}$$

33. The expression $127-73+33-27$ has a -73, a $+33$, and a -27. We circle the expressions that start with 127 and have a -73, a $+33$, and a -27.

(127+33−73−27) 127−27+73−33
127−33+73−27 (127−27+33−73)

34. The expression $136+20+64-36$ has a $+20$, a $+64$, and a -36. We circle the expressions that start with 136 and have a $+20$, a $+64$, and a -36.

136+36−64+20 (136−36+64+20)
136+64−20−36 (136+64−36+20)

35. Subtracting 36 then adding 55 is the same as adding 55 then subtracting 36. So,

$$111-36+55=111 \oplus 55 \ominus 36.$$

Subtracting 45 then adding 132 is the same as adding 132 then subtracting 45. So,

$$168-45+132=168 \oplus 132 \ominus 45.$$

Adding 17, then subtracting 58, then adding 42 is the same as subtracting 58, then adding 42, then adding 17. So,

$$158+17-58+42=158 \ominus 58 \oplus 42 \oplus 17.$$

36. Adding 49 then subtracting 68 is the same as subtracting 68 then adding 49. So,

$$78+49-68=78-\boxed{\mathbf{68}}+\boxed{\mathbf{49}}.$$

Subtracting 44, then subtracting 35, then adding 22 is the same as subtracting 35, then subtracting 44, then adding 22. So,

$$135-44-35+22=135-\boxed{\mathbf{35}}-44+\boxed{\mathbf{22}}.$$

Subtracting 38, then subtracting 12, then adding 9 is the same as adding 9, then subtracting 12, then subtracting 38. So,

$$91-38-12+9=91+\boxed{\mathbf{9}}-\boxed{\mathbf{12}}-38.$$

37. We have

$$\begin{aligned}222+44-222&=222-222+44\\&=0+44\\&=\mathbf{44}.\end{aligned}$$

38. We have

$$\begin{aligned}222-44+222&=222+222-44\\&=444-44\\&=\mathbf{400}.\end{aligned}$$

39. We have

$$\begin{aligned}235+68-35&=235-35+68\\&=200+68\\&=\mathbf{268}.\end{aligned}$$

40. We have

$$\begin{aligned}235-68+35&=235+35-68\\&=270-68\\&=\mathbf{202}.\end{aligned}$$

41. We have

$$\begin{aligned}95+14-30+5&=95+5-30+14\\&=100-30+14\\&=70+14\\&=\mathbf{84}.\end{aligned}$$

42. We have

$$\begin{aligned}88+56-44-25&=88-44+56-25\\&=44+56-25\\&=100-25\\&=\mathbf{75}.\end{aligned}$$

43. We have

$$\begin{aligned}142-34+123-42&=142-42-34+123\\&=100-34+123\\&=66+123\\&=\mathbf{189}.\end{aligned}$$

44. We have

$$\begin{aligned}60+65-30+35&=60-30+65+35\\&=30+65+35\\&=30+100\\&=\mathbf{130}.\end{aligned}$$

45. We have

$$\begin{aligned}235+88-135-57&=235-135+88-57\\&=100+88-57\\&=188-57\\&=\mathbf{131}.\end{aligned}$$

46. We have

$$\begin{aligned}81-37-43+19&=81+19-37-43\\&=100-37-43\\&=100-80\\&=\mathbf{20}.\end{aligned}$$

STRATEGIES

Crosstiles 44-47

47. In the top row, to get from 7 to 15, we add 8. From the given choices, we can only add 8 using +3 and +5.

In the left column, to get from 8 to 15, we add 7. We can only add 7 using +2 and +5.

Since +5 must go in the top row and in the left column, we place +5 in the top-left square.

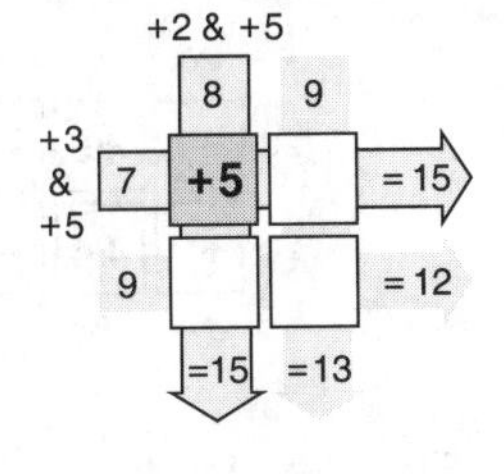

We place +3 to complete the top row and +2 to complete the left column as shown.

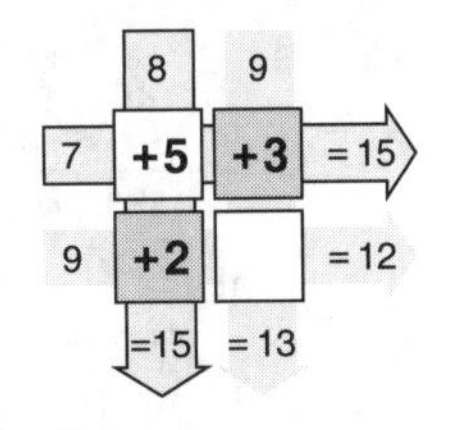

We place the remaining +1 in the last empty square, then check our work as shown.

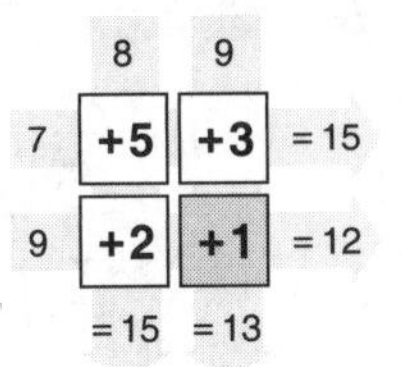

Across:

7+5+3=15. ✓

9+2+1=12. ✓

Down:

8+5+2=15. ✓

9+3+1=13. ✓

48. In the top row, we can only get from 15 to 5 using −2 and −8.

In the left column, we can only get from 21 to 10 using −3 and −8.

Since −8 must go in the top row and in the left column, we place −8 in the top-left square.

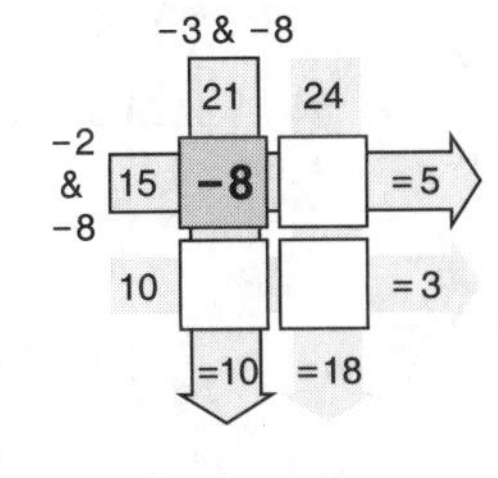

We place −2 to complete the top row and −3 to complete the left column.

Finally, we place −4 in the bottom-right square.

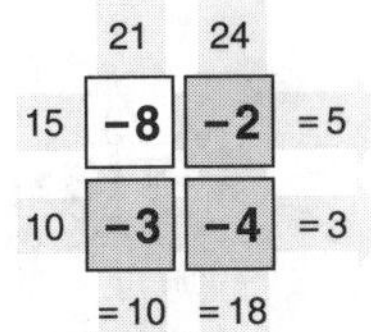

We use the strategies discussed in the previous solutions to solve the Crosstile puzzles that follow.

49. Step 1: Final:

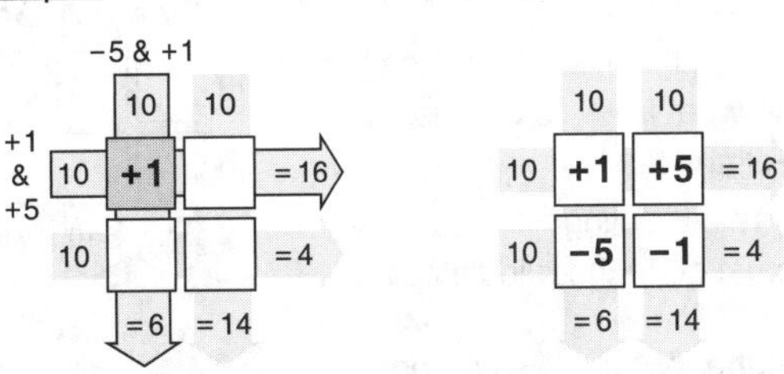

50. Step 1: Final:

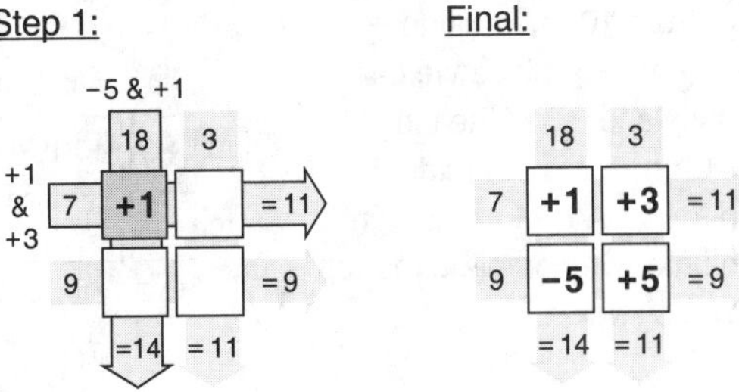

51. Step 1: Final:

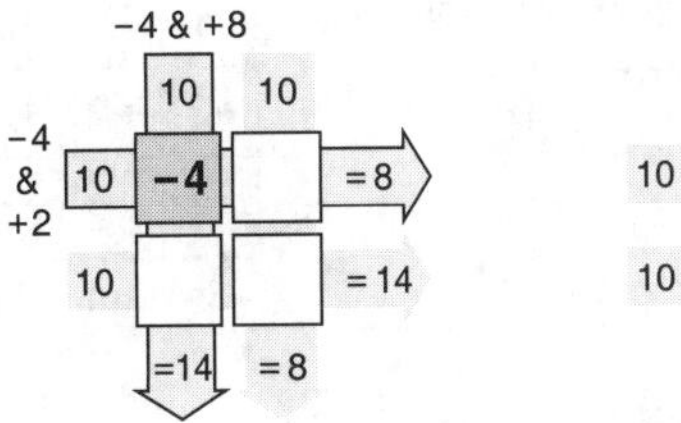

52. Step 1:

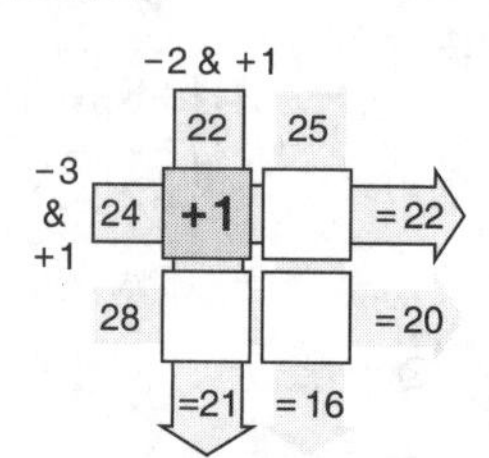

Final:

	22	25	
24	**+1**	**−3**	= 22
28	**−2**	**−6**	= 20
	= 21	= 16	

53. Step 1:

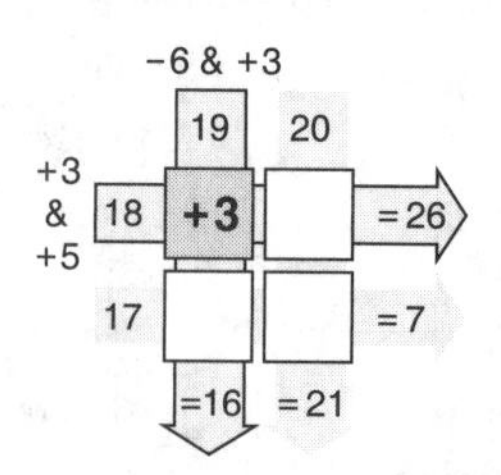

Final:

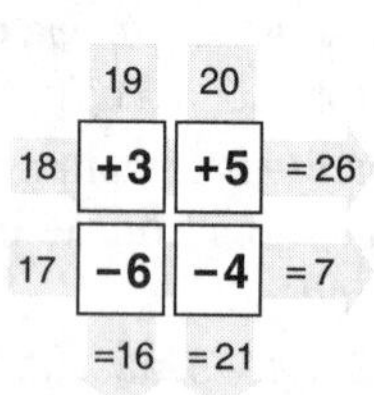

54. Step 1:

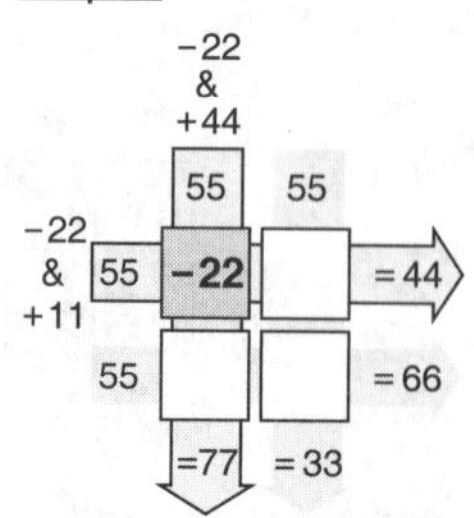

Final:

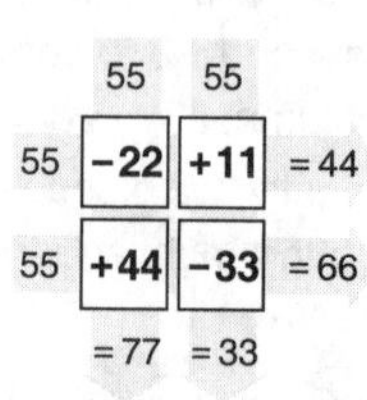

55. In the left column, we can only get from 10 to 12 using −1 and +3.

In the right column, we can only get from 10 to 8 using −3 and +1.

The remaining −2 and +2 must go in the middle column.

	−1 & +3	−2 & +2	−3 & +1	
	10	10	10	
10				= 16
10				= 4
	=12	=10	=8	

In the top row, we can only get from 10 to 16 using +1, +2, and +3. Since we can only place +3 in the left column, +2 in the middle column, and +1 in the right column, we complete the top row as shown.

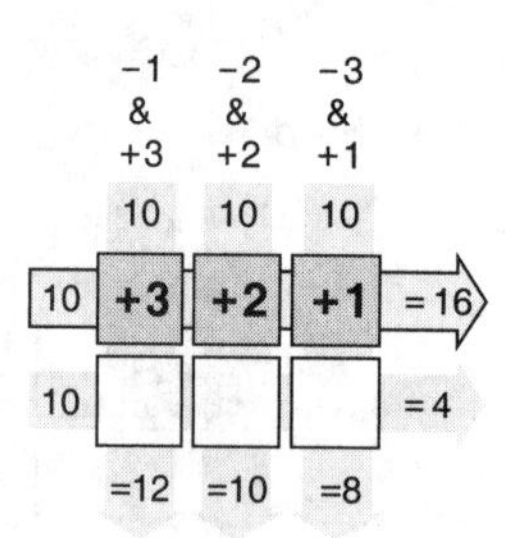

Finally, there is only one way to fill the bottom row.

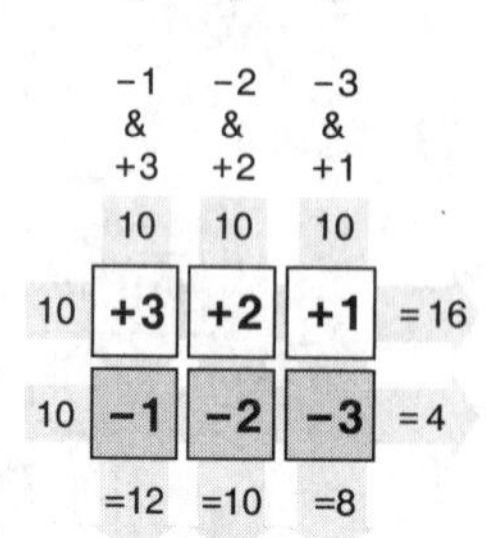

56. Step 1:

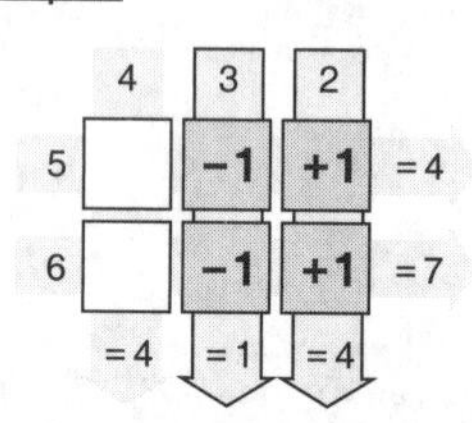

Final:

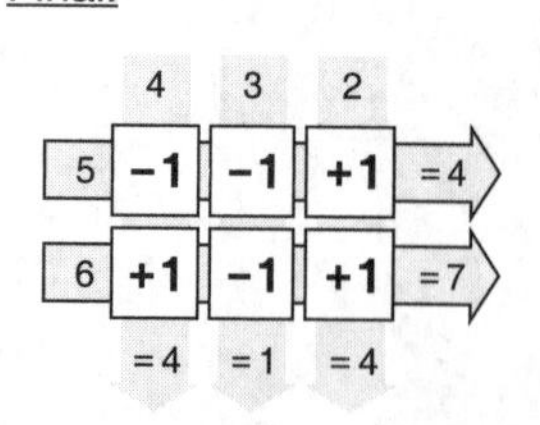

57. In the left column, we can only get from 14 to 20 using +2 and +4.

In the right column, we can only get from 16 to 12 using −1 and −3.

The remaining +6 and −5 must go in the middle column.

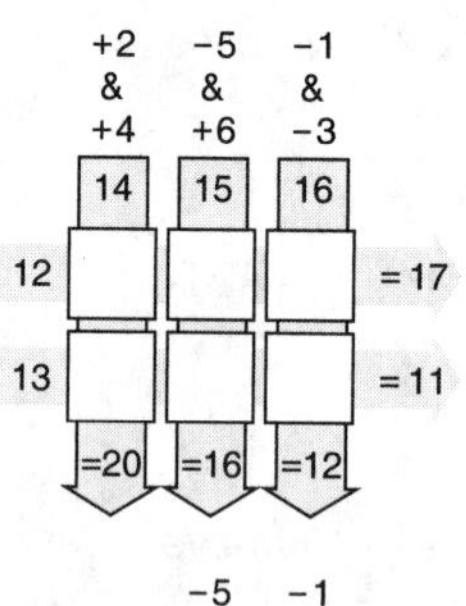

In the left column, if we place +4 above +2, then we cannot complete either row. So, we place +2 above +4.

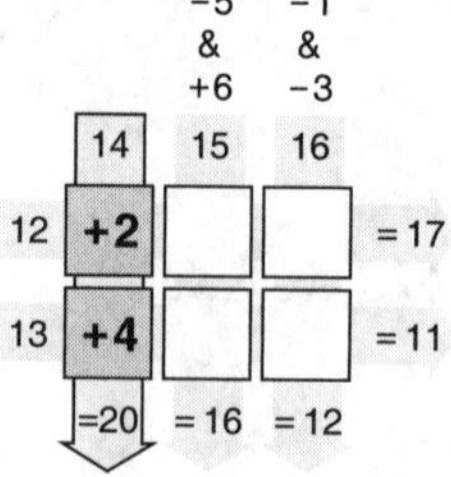

Then, only 13+4−5−1 gives 11 in the bottom row.

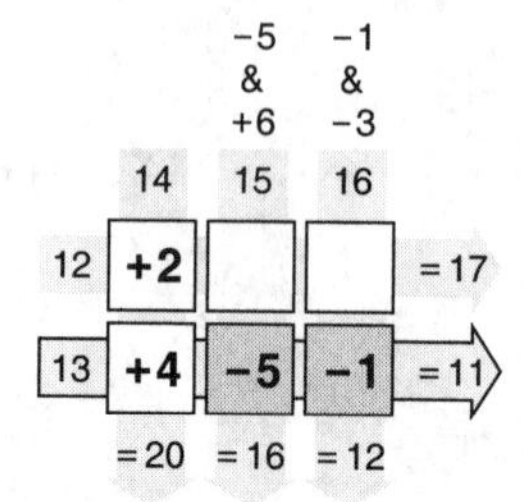

Finally, we place +6 and −3 to complete the middle and right columns.

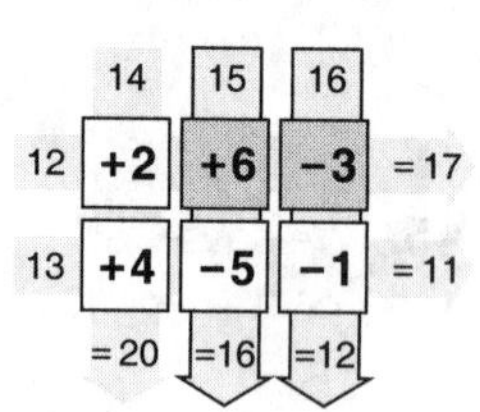

58. Step 1:

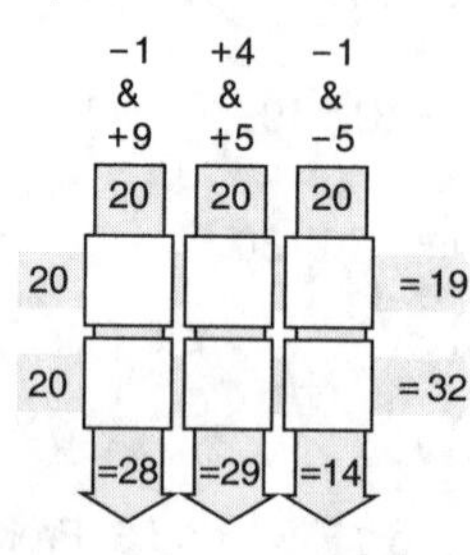

Step 2:

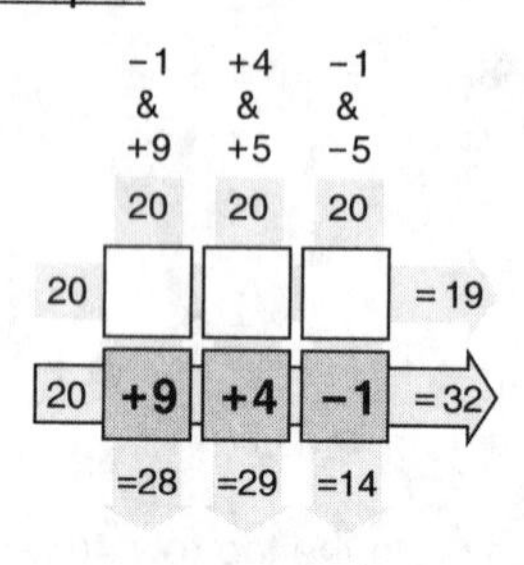

Final:

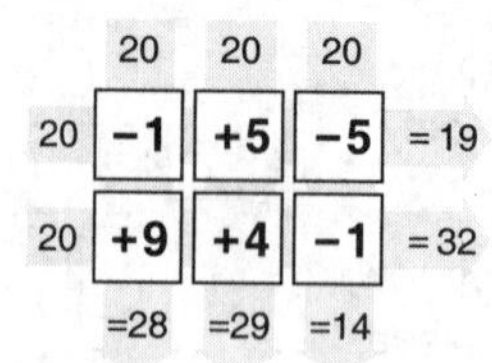

59. Step 1:

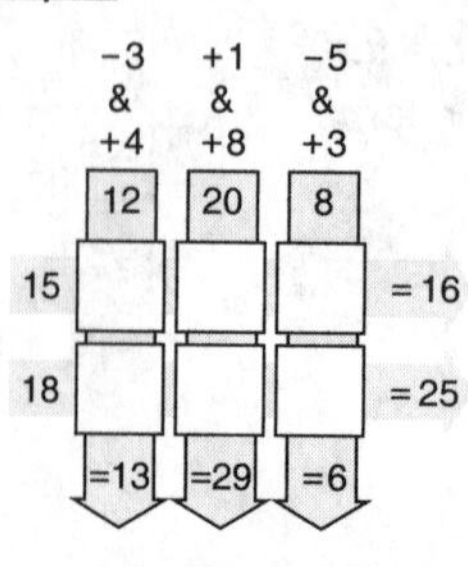

Step 2:

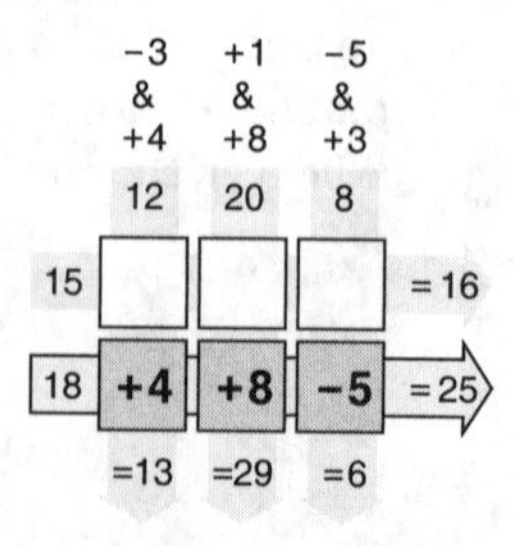

Final:

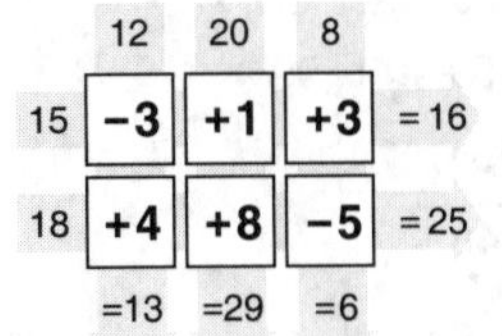

60. Step 1:

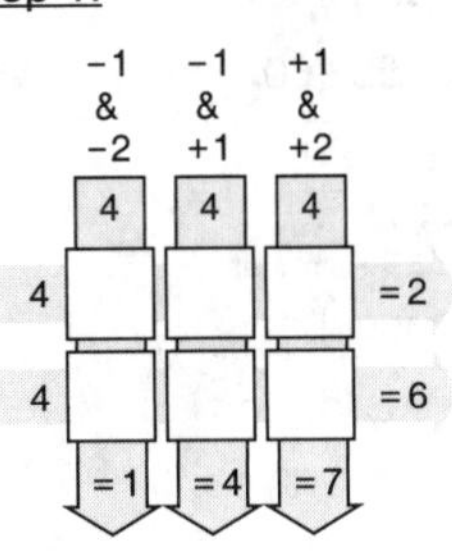

Step 2:

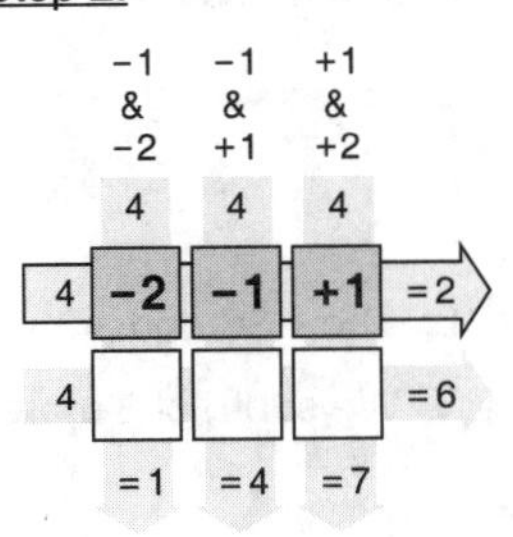

Final:

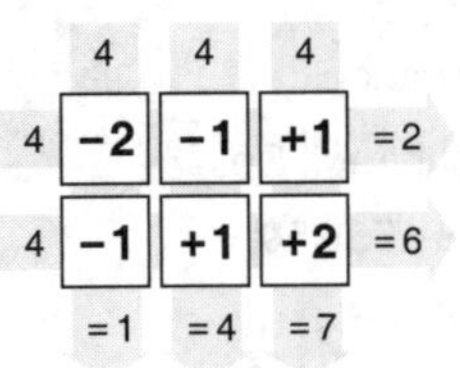

61. Step 1:

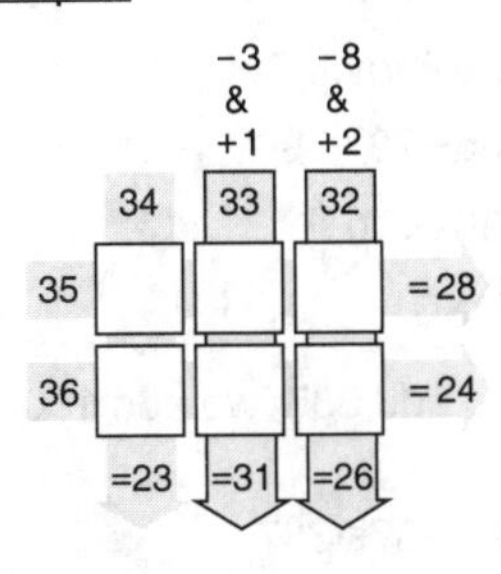

Step 2:

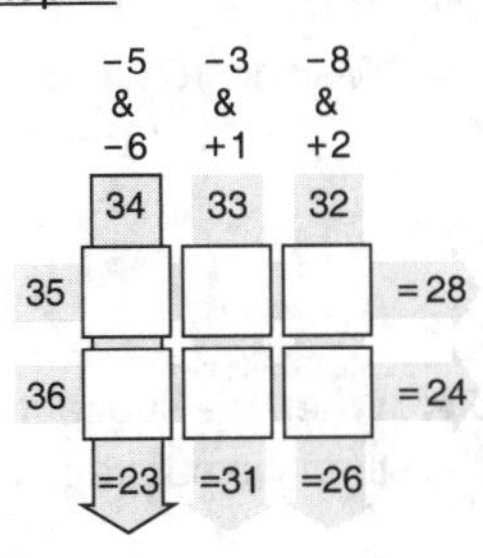

Step 3:

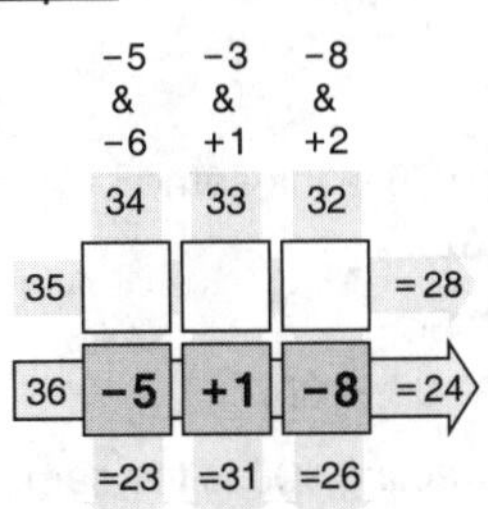

Final:

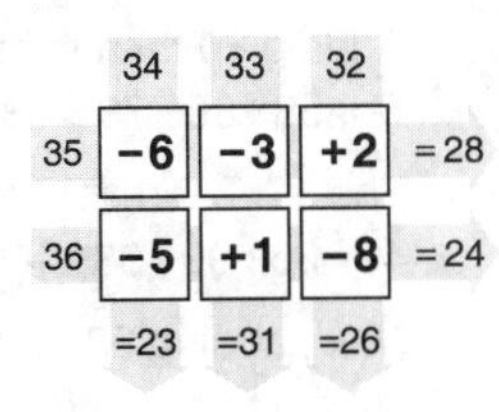

62. Step 1:

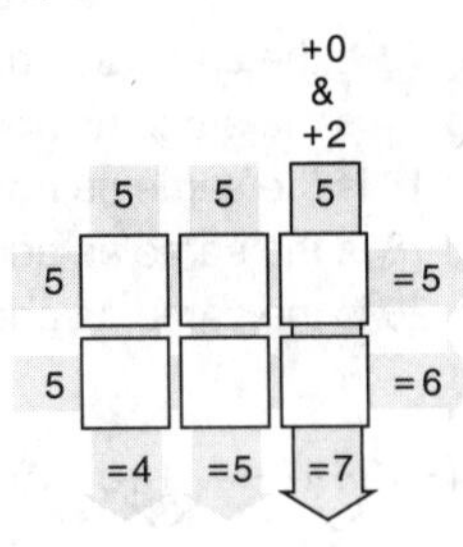

Step 2:

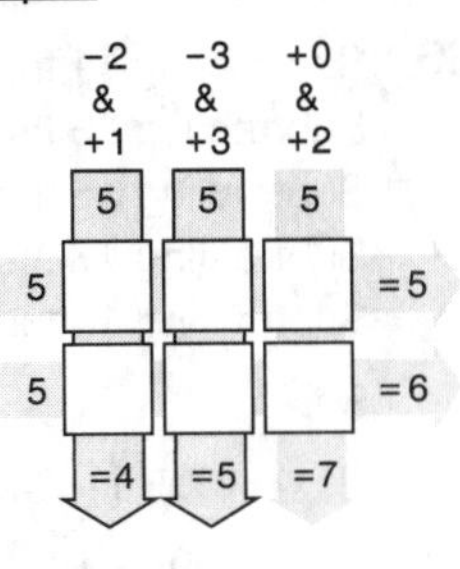

Step 3: Final:

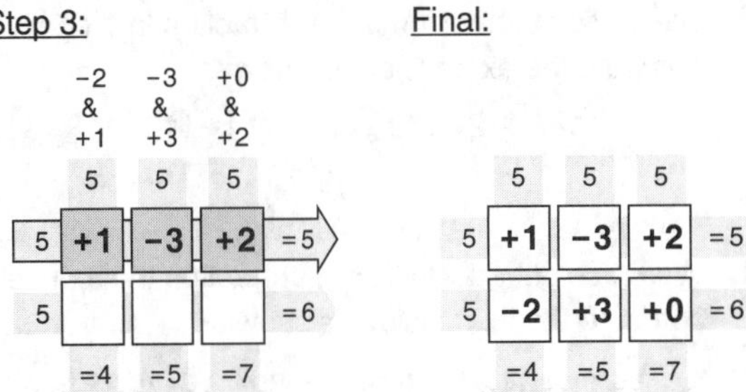

STRATEGIES
Canceling 48–49

63. The elevator is now on floor 15+78−78 = 93−78 = **15**.

— *or* —

An elevator that goes up 78 floors then down 78 floors finishes on the same floor where it started. So, the elevator is on floor **15**.

64. The train has 136+75−36−75 passengers. Having 75 passengers get on and 75 passengers get off does not change the total number of passengers on the train.

So, there are now 136−36 = **100** passengers on the train.

65. Since we are counting the total number of cookies, the type of each cookie does not matter.

Grogg's mom bakes 12, 15, and 24 cookies. Grogg eats 6, 12, and 15 cookies. So there are 12+15+24−6−12−15 cookies left.

Baking 12 cookies and eating 12 leaves 0 cookies. Baking 15 cookies and eating 15 leaves 0 cookies. Baking 24 cookies and eating 6 leaves 24−6 = 18 cookies.

So, there are **18** cookies left.

66. Adding 58 and subtracting 58 is the same as doing nothing, so we have

53+43+58−58 = 53+43+58−58
= 53+ 43
= 96.

67. Subtracting 11 and adding 11 is the same as doing nothing, so we have

19−11+11+7 = 19−11+11+7
= 19+ 7
= 26.

68. Adding 43 and subtracting 43 is the same as doing nothing, so we have

36+43−32−43 = 36+43−32−43
= 36− 32
= 4.

69. We cross out addition and subtraction that cancel, and compute the expression as shown.

13+88+35−88 = 13+88+35−88
= 13+35
= **48**.

70. We cross out addition and subtraction that cancel, and compute the expression as shown.

$$\begin{aligned}114+112-113-112 &= 114+112-113-112\\ &= 114-113\\ &= \mathbf{1}.\end{aligned}$$

71. We cross out addition and subtraction that cancel, and compute the expression as shown.

$$\begin{aligned}80+34-79-34+67 &= 80+34-79-34+67\\ &= 80-79+67\\ &= 1+67\\ &= \mathbf{68}.\end{aligned}$$

72. We cross out addition and subtraction that cancel, and compute the expression as shown.

$$\begin{aligned}99+88+33-88+33 &= 99+88+33-88+33\\ &= 99+33+33\\ &= 99+66\\ &= \mathbf{165}.\end{aligned}$$

73. We cross out addition and subtraction that cancel, and compute the expression as shown.

$$\begin{aligned}&10+9+8-7-8-9+6+8+7+6\\ &= 10+9+8-7-8-9+6+8+7+6\\ &= 10+8-7-8+6+8+7+6\\ &= 10-7+6+8+7+6\\ &= 10+6+8+6\\ &= \mathbf{30}.\end{aligned}$$

74. We cross out addition and subtraction that cancel, and compute the expression as shown.

$$\begin{aligned}&23+32+22-33+23-22+33-32\\ &= 23+32+22-33+23-22+33-32\\ &= 23+22-33+23-22+33\\ &= 23-33+23+33\\ &= 23+23\\ &= \mathbf{46}.\end{aligned}$$

75. There are now $55-36+38=19+38=$ **57** bags of potato chips.

— or —

The deli sold 36 bags of chips, then added 38 more. So, they added $38-36=2$ more bags of chips than they sold. This means that the total number of bags of chips increased by 2.

So, there are now $55+2=$ **57** bags of chips on the rack.

76. Alligophers ate 78 flowers, then 75 flowers grew back. So, the alligophers ate $78-75=3$ more flowers than grew back. This means that the total number of flowers decreased by 3.

So, Blorta now has $120-3=$ **117** flowers in her garden.

77. Gaining 63 points from scoring a clonk then losing 62 points from scoring a bonk is the same as gaining 1 point.

Since Winnie scored three pairs of clonks and bonks, Winnie has $1+1+1=$ **3** points.

78. The +100 and −99 almost cancel.

We add 1 more than we subtract. So,

$$\begin{aligned}&45+100-99\\ &= 45+\boxed{1}\\ &= \boxed{\mathbf{46}}.\end{aligned}$$

79. The −55 and +60 almost cancel.

We add 5 more than we subtract. So,

$$\begin{aligned}&72-55+60\\ &= 72+\boxed{5}\\ &= \boxed{\mathbf{77}}.\end{aligned}$$

80. The −36 and +33 almost cancel.

We subtract 3 more than we add. So,

$$\begin{aligned}&84-36+33\\ &= 84-\boxed{3}\\ &= \boxed{\mathbf{81}}.\end{aligned}$$

81. The +55 and −50 almost cancel.

We add 5 more than we subtract. So,

$$\begin{aligned}&68+55-50\\ &= 68+\boxed{5}\\ &= \boxed{\mathbf{73}}.\end{aligned}$$

82. The −79 and +89 almost cancel.

We add 10 more than we subtract. So,

$$\begin{aligned}&436-79+89\\ &= 436+\boxed{10}\\ &= \boxed{\mathbf{446}}.\end{aligned}$$

83. When we subtract 287 and add 387, we add 100 more than we subtract. So,

$$\begin{aligned}&839-287+387\\ &= 839+\boxed{100}\\ &= \boxed{\mathbf{939}}.\end{aligned}$$

84. Adding 80 then subtracting 76 is the same as adding 4. So, $58+80-76=58+4=$ **62**.

85. The −33 and +33 cancel. So, $49-33+99+33-100=49+99-100$.

Next, adding 99 then subtracting 100 is the same as subtracting 1.

So, we have $49+99-100=49-1=$ **48**.

86. Subtracting 11 then adding 12 is the same as adding 1. Subtracting 13 then adding 14 is the same as adding 1. Subtracting 15 then adding 16 is the same as adding 1. Subtracting 17 then adding 18 is the same as adding 1. Subtracting 19 then adding 20 is the same as adding 1.

So,

$$\begin{aligned}&100-11+12-13+14-15+16-17+18-19+20\\ &= 100\;\;+1\;\;+1\;\;+1\;\;+1\;\;+1\\ &= \mathbf{105}.\end{aligned}$$

STRATEGIES
All At Once — 52

87. Subtracting 8 then subtracting 2 is the same as subtracting 10.

So, 65−8+35−2 is equal to 65+35−10.

65 −8 +35 −2

= 65+35− **10**

= **90**.

88. Subtracting 24 then subtracting 26 is the same as subtracting 50.

So, 53−24−26+38 is equal to 53−50+38.

53 −24 −26 +38

= 53− **50** +38

= **41**.

89. Adding 47 then adding 53 is the same as adding 100.

So, 125+47−123+53 is equal to 125−123+100.

125 +47 −123 +53

= 125−123+ **100**

= **102**.

90. Subtracting 12 then subtracting 18 is the same as subtracting 30.

So, 50−12+20−18 is equal to 50+20−30.

50 −12 +20 −18

= 50+20− **30**

= **40**.

91. Subtracting 38 then subtracting 122 is the same as subtracting 160.

So, 200−38+700−122 is equal to 200+700−160.

200 −38 +700 −122

= 200+700− **160**

= **740**.

92. Adding 15 then adding 15 is the same as adding 30.

Subtracting 6 then subtracting 6 is the same as subtracting 12.

So, 19+15+15−6−6 is equal to 19+30−12.

19 +15 +15 −6 −6

= 19+ **30** − **12**

= **37**.

93. Subtracting 19 then subtracting 21 is the same as subtracting 40.

Adding 33 then adding 17 is the same as adding 50.

So, 134−19+33−21+17 is equal to 134−40+50.

134 −19 +33 −21 +17

= 134− **40** + **50**

= **144**.

94. Adding 81 then adding 9 is the same as adding 90.

Subtracting 11 then subtracting 33 is the same as subtracting 44.

So, 7+81−11−33+9 is equal to 7+90−44.

7 +81 −11 −33 +9

= 7+ **90** − **44**

= **53**.

95. Subtracting 19 then subtracting 11 is the same as subtracting 30.

Adding 22 then adding 23 is the same as adding 45.

So, 80−19−11+22+23 is equal to 80−30+45.

80 −19 −11 +22 +23

= 80− **30** + **45**

= **95**.

STRATEGIES
Missing Numbers — 53

96. Starting with 28, we add 19 then subtract a number to get 28. So, the +19 and −☐ must cancel.

Since +19 and −19 cancel, the number in the box is 19.

28+19− **19** = 28.

97. 789+☐−789 is equal to 789−789+☐.

Since 789−789 is 0, we have 0+☐ = 987.

Since 0+987 = 987, the number in the box is 987.

789+ **987** −789 = 987.

98. Starting with 64, we subtract 27 then add ☐ to get 65.

65 is 1 more than 64. So, we add 1 more than we subtract. Since we subtract 27, the number in the box is 27+1 = 28.

64−27+ **28** = 65.

99. 113−☐+113 is equal to 113+113−☐.

Since 113+113 = 226, we have 226−☐ = 200.

Since 226−26 = 200, the number in the box is 26.

113− **26** +113 = 200.

100. Starting with 226, we subtract 77 then add ☐ to get 224.

224 is 2 less than 226. So, we subtract 2 more than we add. Since we subtract 77, the number in the box is 75.

226−77+ **75** = 224.

101. 75−28+75−☐ is equal to 75+75−28−☐.

Since 75+75 = 150, we have 150−28−☐ = 100.

Starting with 150, we must subtract 50 to get to 100.

Subtracting 28 then subtracting 22 is the same as subtracting 50 all at once. So, the number in the box is 22.

75−28+75−**22** = 100.

102. Starting with 77, we add and subtract some numbers to get 77. So, +144, −12, +☐, and −145 must all cancel.

Adding 144 then subtracting 145 is the same as subtracting 1.

So, subtracting 12 then adding ☐ must be the same as adding 1.

So, the number in the box is 13.

77+144−12+**13**−145 = 77.

103. Starting with 200, we add and subtract some numbers to get 203. So, we must add a total of 3.

Subtracting 26 then adding 27 is the same as adding 1.

So, subtracting ☐ then adding 84 must be the same as adding 2 more.

So, the number in the box is 82.

200−28−**82**+27+84 = 203.

STRATEGIES
Equation Paths 54–56

104. To get from 15 to 28, we add a total of 13. To add 13, our path must cross one +10 and three +1's.

15	+1	+1
+1	+10	+1
+10	+1	+10 = 28

105. To get from 14 to 70, we add a total of 56. To add 56, our path must cross +10, +20, and +30. The path must also cross −4 at the end of the path.

14	−1	−2
+10	−2	−3
+20	+30	−4 = 70

106. To get from 100 to 90, we subtract a total of 10. We can do this by crossing two −5's. Any other numbers in our path must cancel.

100	+17	−31
+3	−5	+3
+31	−17	−5 = 90

107. To get from 26 to 32, we add a total of 6. We can do this by crossing two +3's. The other numbers in our path must cancel.

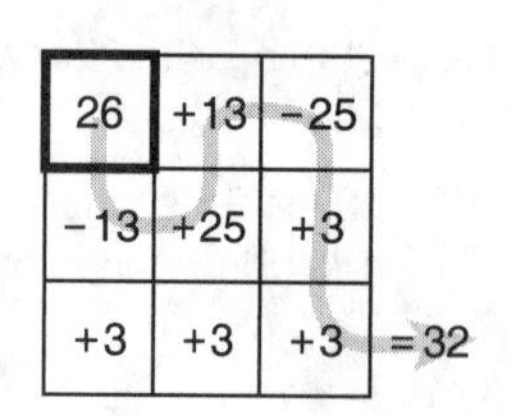

26	+13	−25
−13	+25	+3
+3	+3	+3 = 32

We use the strategies discussed in the previous solutions to solve the Equation Path puzzles that follow.

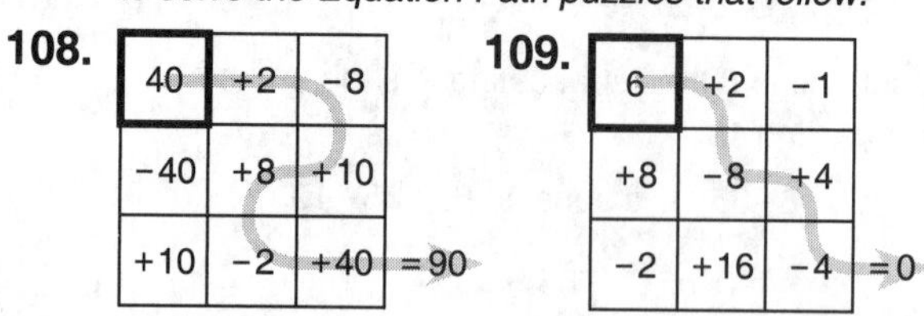

108.

40	+2	−8
−40	+8	+10
+10	−2	+40 = 90

109.

6	+2	−1
+8	−8	+4
−2	+16	−4 = 0

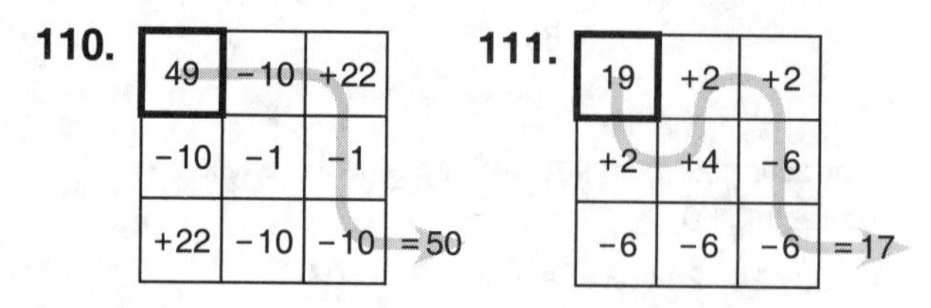

110.

49	−10	+22
−10	−1	−1
+22	−10	−10 = 50

111.

19	+2	+2
+2	+4	−6
−6	−6	−6 = 17

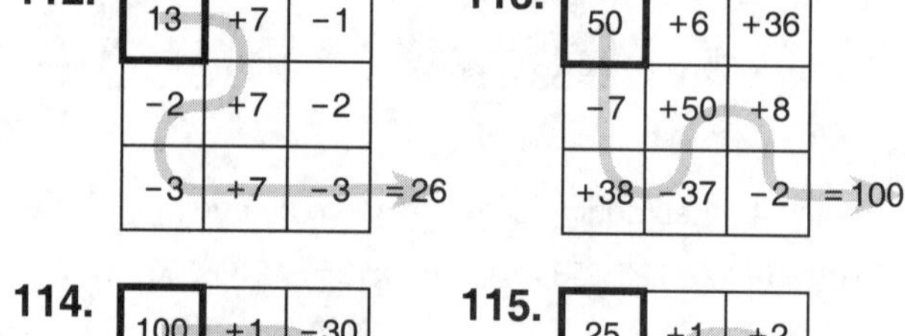

112.

13	+7	−1
−2	+7	−2
−3	+7	−3 = 26

113.

50	+6	+36
−7	+50	+8
+38	−37	−2 = 100

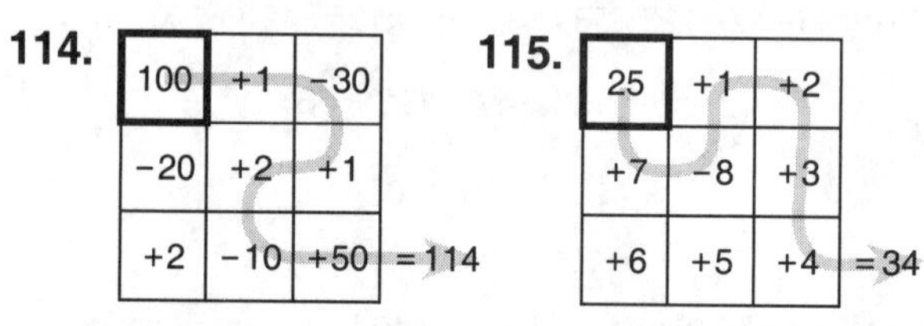

114.

100	+1	−30
−20	+2	+1
+2	−10	+50 = 114

115.

25	+1	+2
+7	−8	+3
+6	+5	+4 = 34

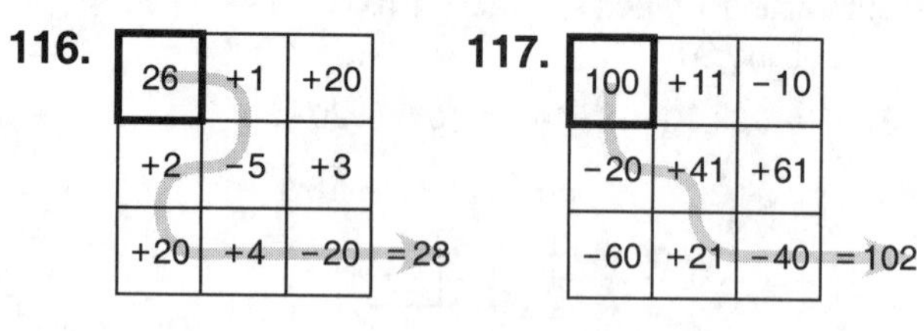

116.

26	+1	+20
+2	−5	+3
+20	+4	−20 = 28

117.

100	+11	−10
−20	+41	+61
−60	+21	−40 = 102

STRATEGIES
Parentheses Review 57

118. 128−(12+16)
= 128−28
= **100**.

119. 55−(61−30)+26
= 55−31+26
= 24+26
= **50**.

120. 50−(11+14−5)
= 50−(25−5)
= 50−20
= **30**.

121. 99−(66−33)−(66−33)
= 99−33−33
= 66−33
= **33**.

122. 27−(30−(9−5)−6)
= 27−(30−4−6)
= 27−(26−6)
= 27−20
= **7**.

123. 99−((66−33)−(66−33))
= 99−(33−33)
= 99−0
= **99**.

124. We check ways we can place parentheses in $22-7-6+5-4$.

There are three ways to group two numbers that change what we evaluate first:

$22-(7-6)+5-4$ gives $22-1+5-4=22$.
$22-7-(6+5)-4$ gives $22-7-11-4=0$.
$22-7-6+(5-4)$ gives $22-7-6+1=10$.

There are two ways to group three numbers that change what we evaluate first:

$22-(7-6+5)-4$ gives $22-6-4=12$.
$22-7-(6+5-4)$ gives $22-7-7=8$.

There is one way to group four numbers that changes what we evaluate first:

$22-(7-6+5-4)$ gives $22-2=20$.

Of our choices, only $22-7-\mathbf{(6+5)}-4=0$ makes a true equation.

STRATEGIES
Parentheses Word Problems 58–59

125. Sam swam a total of $17+17$ laps. Myles swam a total of $15+15$ laps. So, Sam swam $(17+17)-(15+15)$ more laps than Myles.

In the morning, Sam swam $17-15$ more laps than Myles. In the afternoon, Sam swam $17-15$ more laps than Myles. So, Sam swam $(17-15)+(17-15)$ more laps than Myles.

We circle the two expressions that show how many more laps Sam swam than Myles.

$(17-15)-(17-15)$ $(17+17)+(15+15)$
(circled) $(17+17)-(15+15)$ **(circled)** $(17-15)+(17-15)$

126. $(17-15)+(17-15)=2+2=4$.

So, Sam swam **4** more laps than Myles.

127. The Merkel brothers weigh a total of $164+138$ pounds. The Samson sisters weigh a total of $161+133$ pounds. So, the Merkel brothers weigh $(164+138)-(161+133)$ pounds more than the Samson sisters.

The heavier Merkel brother weighs $164-161$ pounds more than the heavier Samson sister. The lighter Merkel brother weighs $138-133$ pounds more than the lighter Samson sister. So, the Merkel brothers weigh $(164-161)+(138-133)$ pounds more than the Samson sisters.

We circle the two expressions that show how many more pounds the Merkel brothers weigh than the Samson sisters.

(circled) $(164+138)-(161+133)$ $(164+161)-(138+133)$
$(164+138)+(161+133)$ **(circled)** $(164-161)+(138-133)$

128. $(164-161)+(138-133)=3+5=8$.

So, the Merkel brothers weigh **8** pounds more than the Samson sisters.

129. Alex has $35+23+14$ candies.
Lizzie has $30+21+12$ candies.

So, Alex has $(35+23+14)-(30+21+12)$ more candies than Lizzie.

Alex has $35-30$ more yellow candies than Lizzie.
Alex has $23-21$ more red candies than Lizzie.
Alex has $14-12$ more green candies than Lizzie.

So, Alex has $(35-30)+(23-21)+(14-12)$ more candies than Lizzie.

We circle the two expressions that show how many more candies Alex has than Lizzie.

$(35+23+14)+(30+21+12)$ **(circled)** $(35+23+14)-(30+21+12)$
(circled) $(35-30)+(23-21)+(14-12)$ $(35+30)-(23+21)-(14+12)$

130. $(35-30)+(23-21)+(14-12)=5+2+2=9$.

So, Alex has **9** more candies than Lizzie.

131. After 4 days, Grogg has made $15+15+15+15$ waffles and eaten $11+11+11+11$ waffles. So, Grogg has $(15+15+15+15)-(11+11+11+11)$ frozen waffles.

Each day, Grogg freezes $15-11$ waffles. So, after 4 days, Grogg has $(15-11)+(15-11)+(15-11)+(15-11)$ frozen waffles.

We circle the two expressions that show how many frozen waffles Grogg has after 4 days.

$(15+15+15+15)-(4+4+4+4)$ **(circled)** $(15+15+15+15)-(11+11+11+11)$
$(15-4)+(15-4)+(15-4)+(15-4)$ **(circled)** $(15-11)+(15-11)+(15-11)+(15-11)$

132. $(15-11)+(15-11)+(15-11)+(15-11)=4+4+4+4=16$.

So, Grogg has **16** frozen waffles after 4 days.

STRATEGIES
Parentheses 60–61

133. Subtracting any expression from itself equals 0. So,

$$(38+57)-(38+57)=(38+57)-(38+57)$$
$$=\mathbf{0}.$$

134. Since we can add numbers in any order, $(23+19)$ is equal to $(19+23)$. So, subtracting $(23+19)$ from $(19+23)$ equals 0. We have

$$(19+23)-(23+19)=(19+23)-(23+19)$$
$$=\mathbf{0}.$$

135. Adding $(9+5)$ then subtracting $(9+5)$ is the same as doing nothing. So,

$$4+(9+5)-(9+5)=4+(9+5)-(9+5)$$
$$=\mathbf{4}.$$

136. Since we can add numbers in any order, $(7+6)$ is equal to $(6+7)$. So, subtracting $(6+7)$ from $(7+6)$ equals 0, and we have

$$(7+6)-(6+7)+8=(7+6)-(6+7)+8$$
$$=\mathbf{8}.$$

137. Adding 7 then subtracting 7 is the same as doing nothing. So,

$$(8+3)+7+(8+3)-7=(8+3)+7+(8+3)-7$$
$$=11+11$$
$$=\mathbf{22}.$$

138. Since we can add numbers in any order, (7+11) is equal to (11+7). So, subtracting (11+7) then adding (7+11) is the same as doing nothing, and we have

$$\begin{aligned}3-(11+7)+(7+11)+3 &= 3-(11+7)+(7+11)+3\\ &= 3+3\\ &= \mathbf{6}.\end{aligned}$$

139. Since we can add numbers in any order, (9+8+7) is equal to (7+8+9). So, adding (7+8+9) then subtracting (9+8+7) is the same as doing nothing, and we have

$$\begin{aligned}6+(7+8+9)+10-(9+8+7) &= 6+(7+8+9)+10-(9+8+7)\\ &= 6+10\\ &= \mathbf{16}.\end{aligned}$$

140. Since we can add numbers in any order, (110+80) is equal to (80+110). So, subtracting (110+80) from (80+110) equals 0, and we have

$$\begin{aligned}(80+110)+50-(110+80)+(110-80) &= 50+(110-80)\\ &= 50+30\\ &= \mathbf{80}.\end{aligned}$$

141. 30 is 1 more than 29. So, (30+30) is 1+1=2 more than (29+29).

So, (30+30)−(29+29)=**2**.

142. 86 is 1 more than 85, and 85 is 1 more than 84. So, (86+85) is 1+1=2 more than (85+84).

So, (86+85)−(85+84)=**2**.

— *or* —

86 is 2 more than 84. So, (86+85) is 2 more than (85+84).

So, (86+85)−(85+84)=**2**.

143. 120 is 20 more than 100, and 121 is 100 more than 21. So, (120+121) is 20+100=120 more than (100+21).

So, (120+121)−(100+21)=**120**.

— *or* —

(100+21)=121, and (120+121) is 120 more than 121.

So, (120+121)−(100+21)=(120+121)−(121)=**120**.

144. 9 is 3 more than 6.
8 is 3 more than 5
7 is 3 more than 4.

So, (9+8+7) is 3+3+3=9 more than (6+5+4).

So, (9+8+7)−(6+5+4)=**9**.

145. Since we can add numbers in any order, (73+75+77) is the same as (77+75+73).

177 is 100 more than 77.
175 is 100 more than 75.
173 is 100 more than 73.

So, (177+175+173) is 100+100+100=300 more than (77+75+73).

So, (177+175+173)−(73+75+77)=**300**.

146. Taking away 14, 15, and 16 from (13+14+15+16+17) leaves 13 and 17.

So, (13+14+15+16+17)−(14+15+16)=13+17=**30**.

147. 82 is 1 more than 81.
84 is 1 more than 83.
86 is 1 more than 85.
88 is 1 more than 87.
90 is 1 more than 89.

So, (82+84+86+88+90) is 1+1+1+1+1=5 more than (81+83+85+87+89).

So, (82+84+86+88+90)−(81+83+85+87+89)=**5**.

148. Since 4 is 2 more than 2, we are skip-counting by 2's.

To skip-count by 2's, we add 2 over and over.

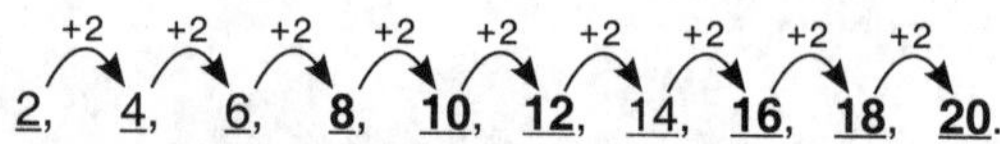

149. We skip-count by 5's to complete the pattern.

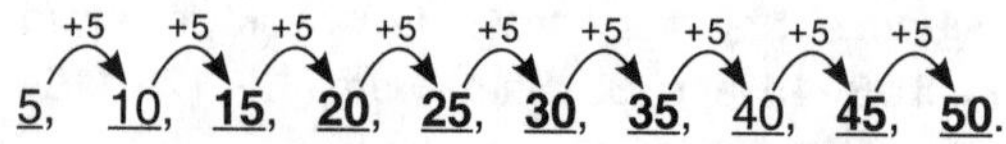

150. We skip-count by 20's to complete the pattern.

+20 +20 +20 +20 +20 +20 +20 +20 +20

20, 40, **60**, **80**, **100**, **120**, **140**, **160**, **180**, **200**.

151. We skip-count by 3's to complete the pattern.

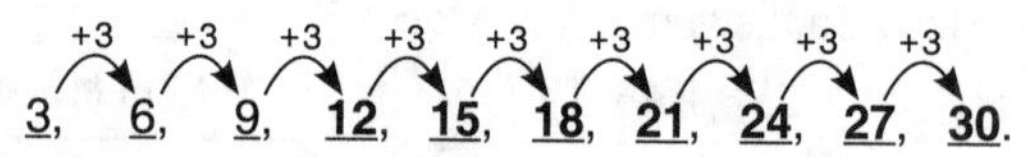

152. We skip-count by 4's to complete the pattern.

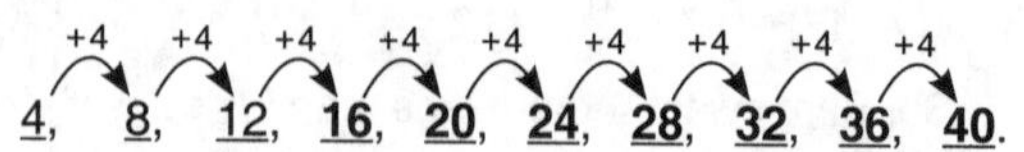

153. We skip-count by 11's to complete the pattern to the right of 77.

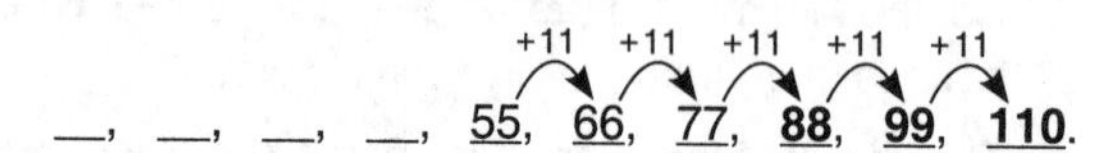

Since the number to the right of 55 is 11 *more* than 55, the number to the left of 55 is 11 *less* than 55. We continue this pattern to the left.

−11 −11 −11 −11

11, **22**, **33**, **44**, 55, 66, 77, **88**, **99**, **110**.

154. Since 198 is 99 more than 99, we are skip-counting by 99's.

To skip-count by 99's, we add 99 over and over. Since 99=100−1, adding 99 is the same as adding 100 then taking away 1.

+99 +99 +99 +99 +99 +99 +99 +99 +99

99, 198, **297**, **396**, **495**, **594**, **693**, **792**, **891**, **990**.

155. There are 2 dots in each group, so we skip-count by 2's to count the dots.

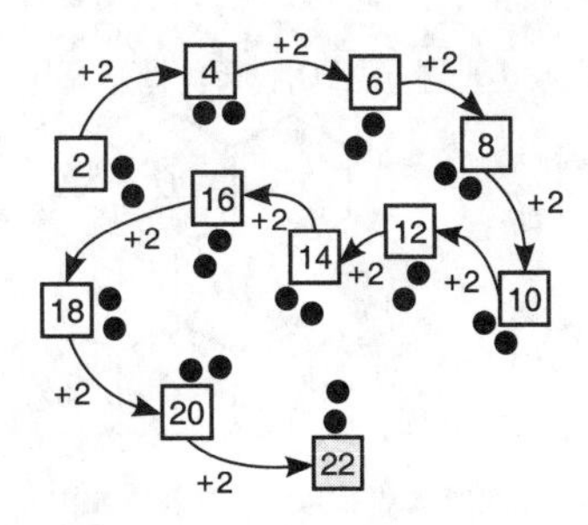

So, there are **22** dots.

156. There are 10 dots in each group, so we skip-count by 10's to count the dots.

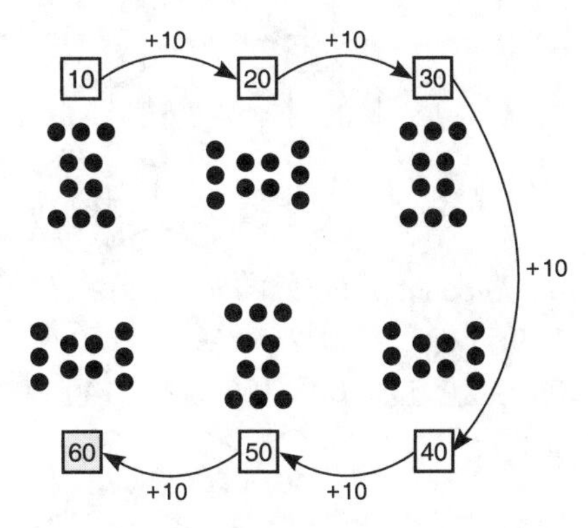

So, there are **60** dots.

157. There are 3 dots in each group, so we skip-count by 3's to count the dots.

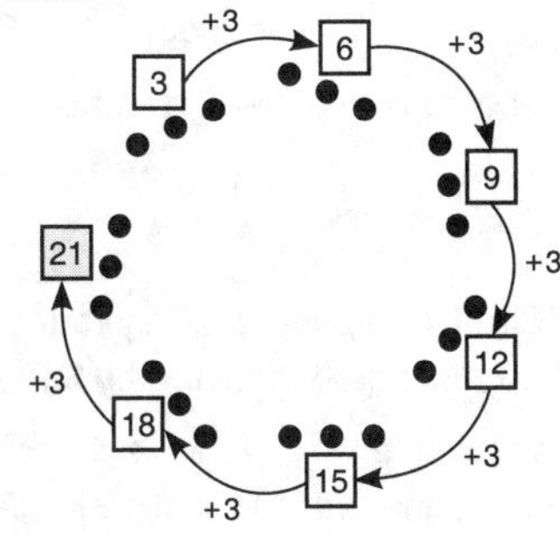

So, there are **21** dots.

158. There are 5 dots in each group, so we skip-count by 5's to count the dots.

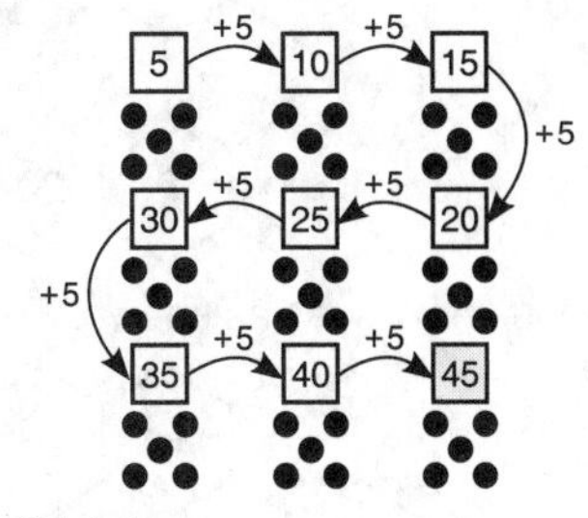

So, there are **45** dots.

159. There are 4 dots in each group, so we skip-count by 4's to count the dots.

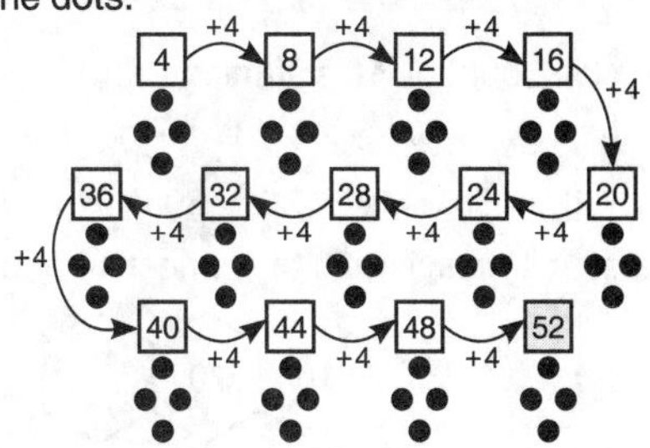

So, there are **52** dots.

160. There are 11 dots in each group, so we skip-count by 11's to count the dots.

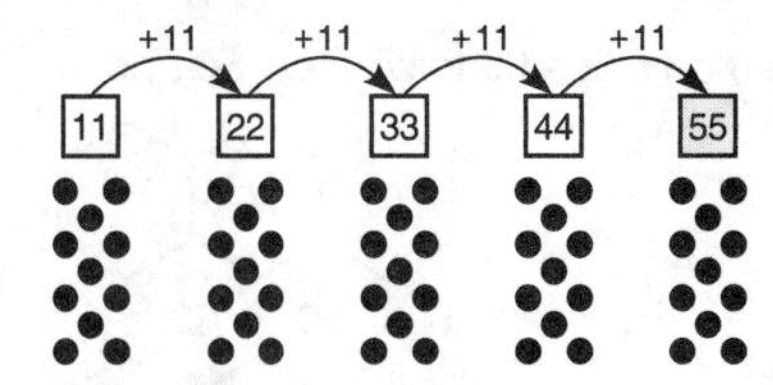

So, there are **55** dots.

161. We can circle groups of 4 dots, then skip-count by 4's to count the dots.

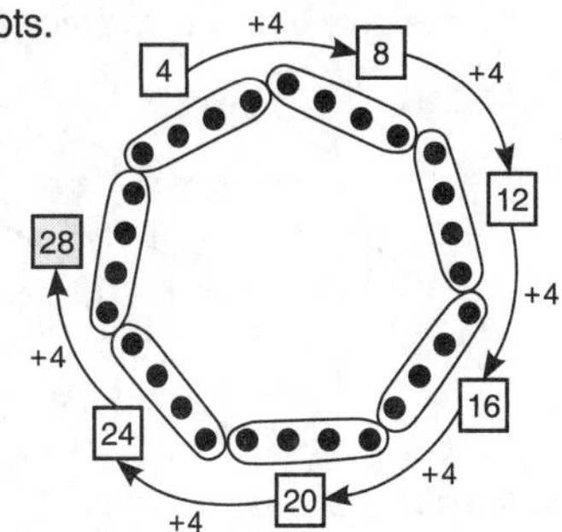

So, there are **28** dots.

162. We can circle groups of 5 dots, then skip-count by 5's to count the dots.

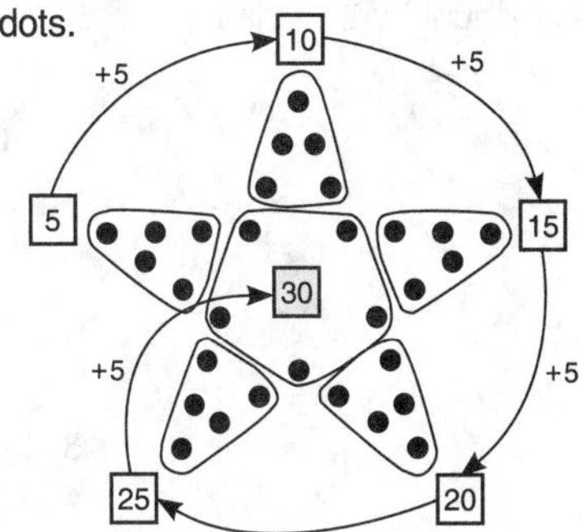

So, there are **30** dots.

— or —

We can circle groups of 3 dots, then skip-count by 3's to count the dots.

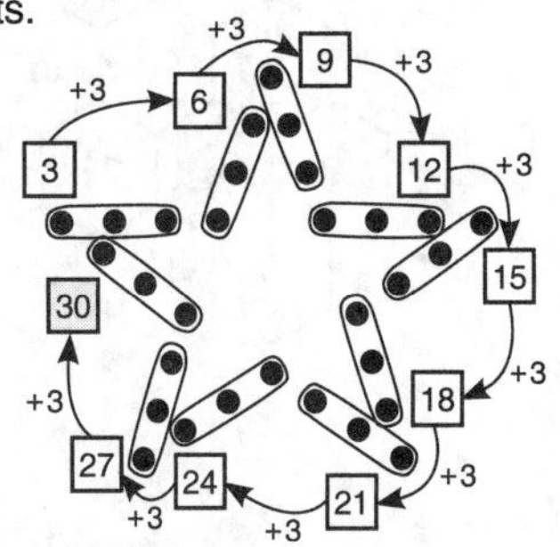

So, there are **30** dots.

STRATEGIES
Honeycomb Paths 64–67

163. We skip-count by 10's starting with 10:

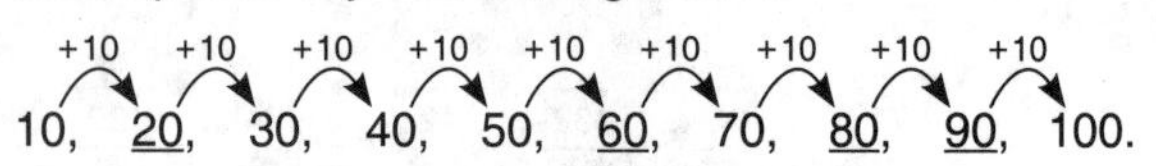

We start by placing 20 to connect 10 and 30.

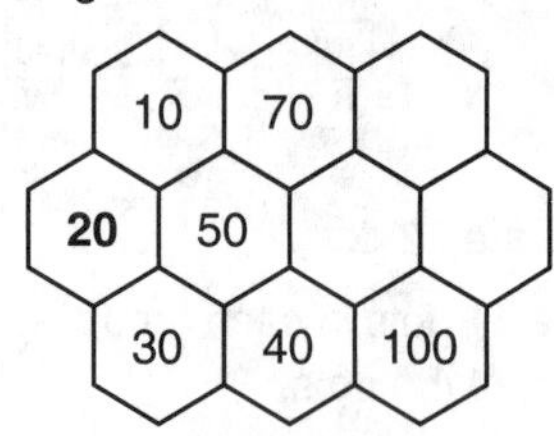

We then place 60 to connect 50 and 70.

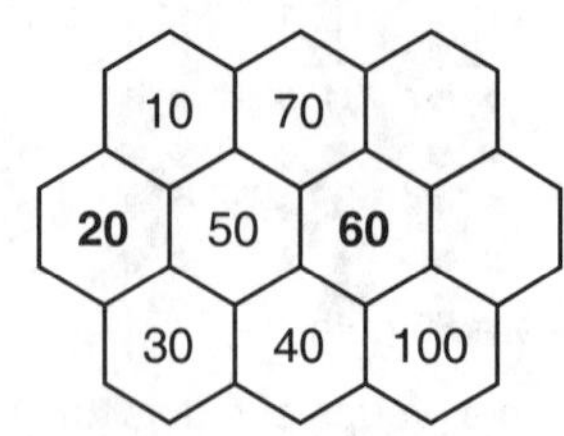

Finally, there is only one way to place 80 and 90 to connect 70 and 100.

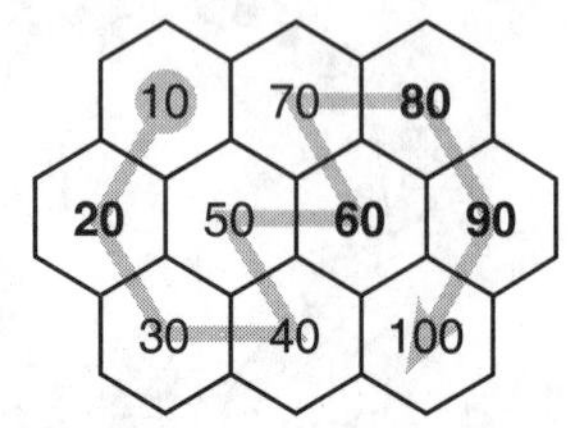

We use the strategies discussed in the previous solution to solve the Skip-Counting Honeycomb Path puzzles that follow.

164. We skip-count by 2's starting with 2:

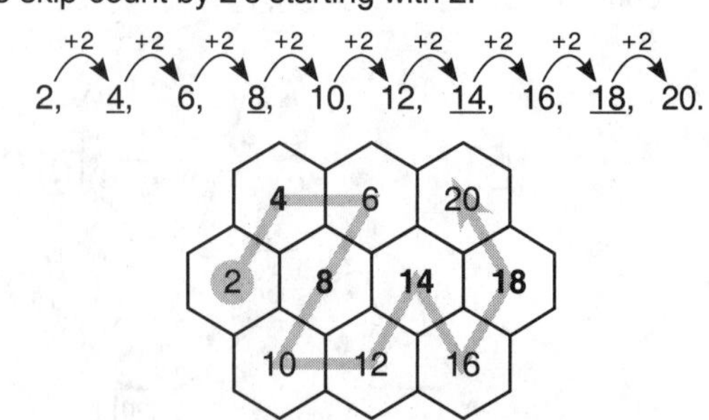

165. We skip-count by 3's starting with 3:

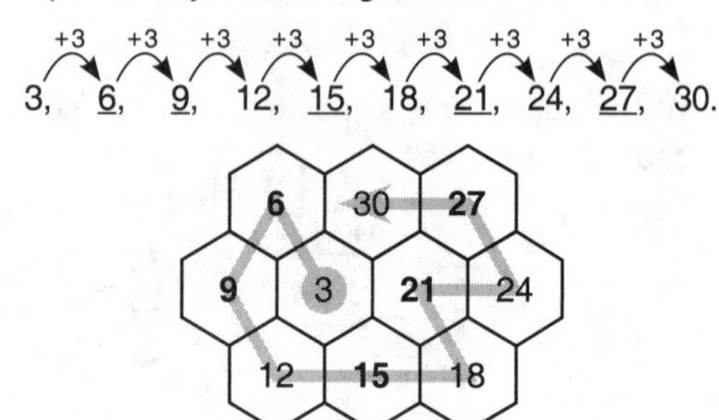

166. We skip-count by 5's starting with 5:

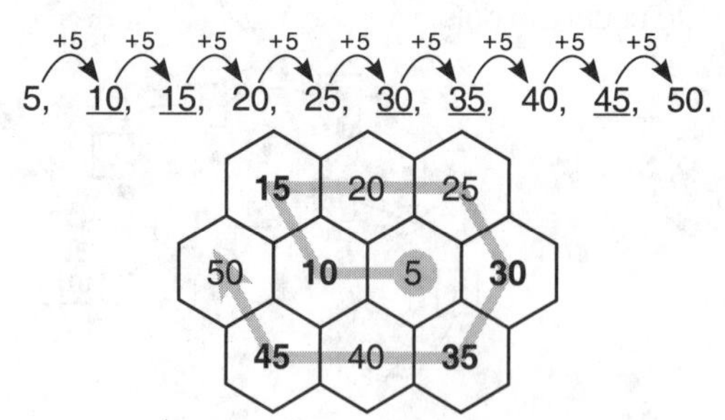

167. We skip-count by 4's starting with 4:

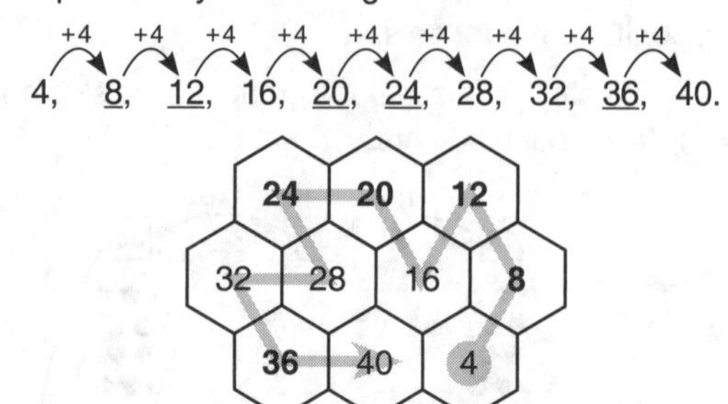

168. We skip-count by 20's starting with 20:

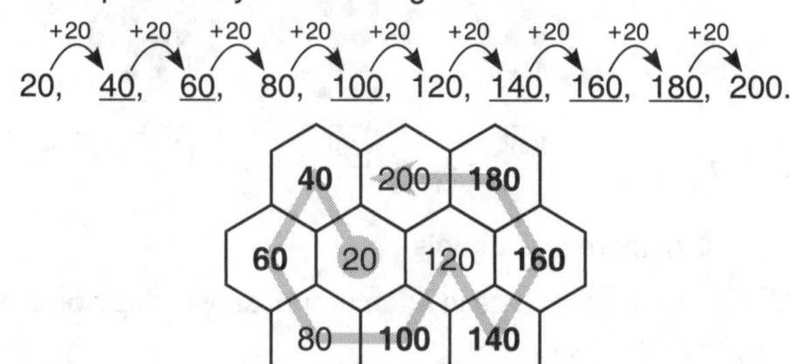

169. We skip-count by 2's starting with 2:

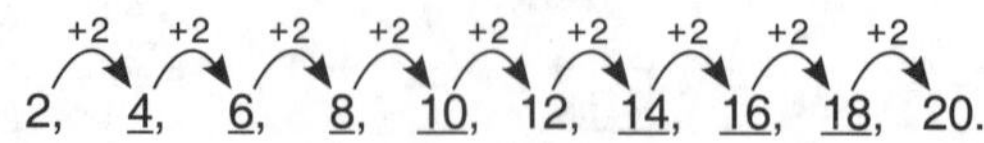

2 and 12 must be connected by four empty hexagons. 12 and 20 must be connected by three empty hexagons.

There is only one way to connect 2 to 12 with four empty hexagons *and* connect 12 to 30 with three empty hexagons, as shown below.

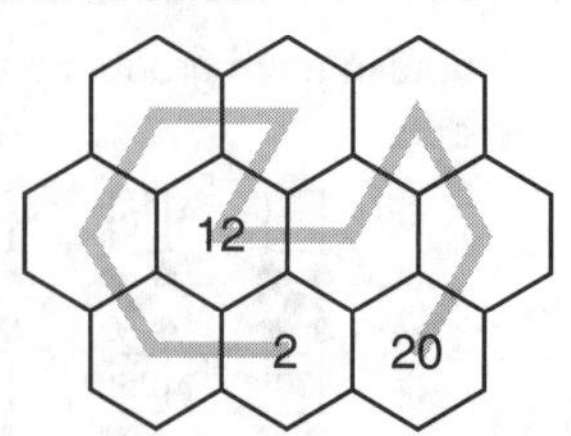

We fill those hexagons as shown.

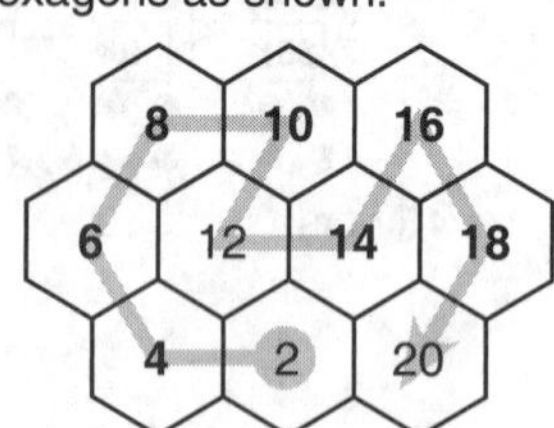

170. We skip-count by 50's starting with 50:

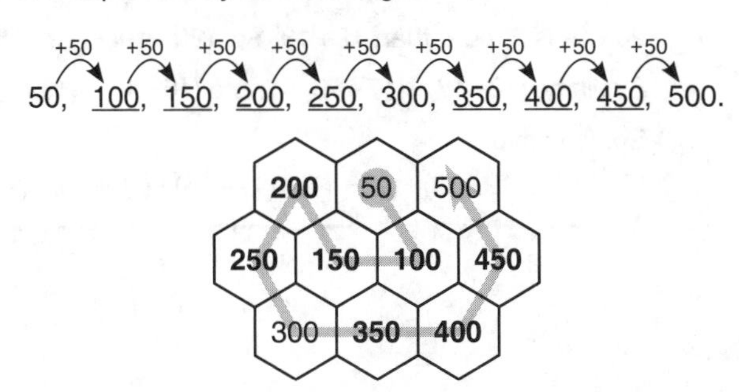

171. We skip-count by 11's starting with 11:

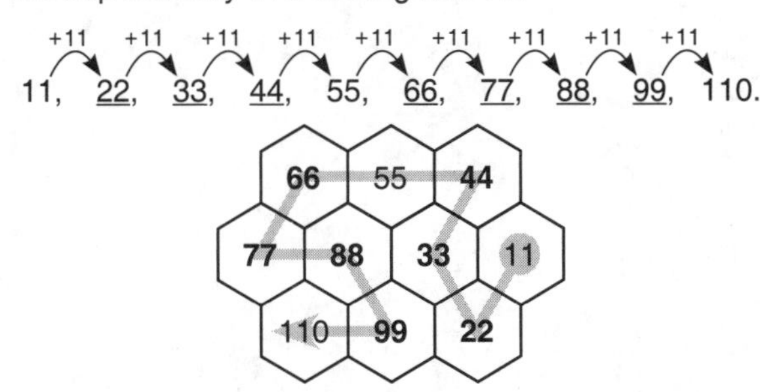

172. We skip-count by 9's starting with 9:

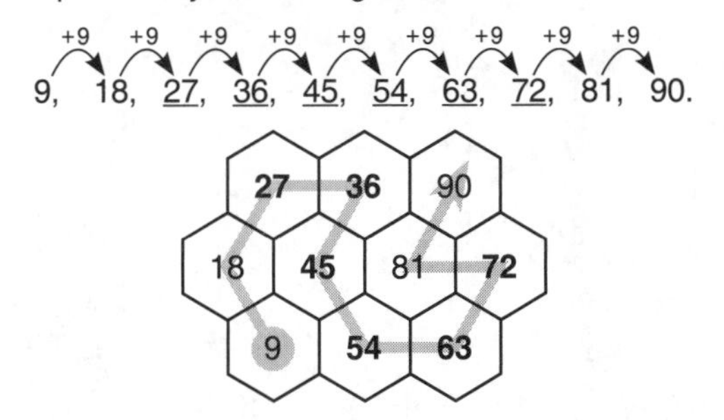

173. **174.**

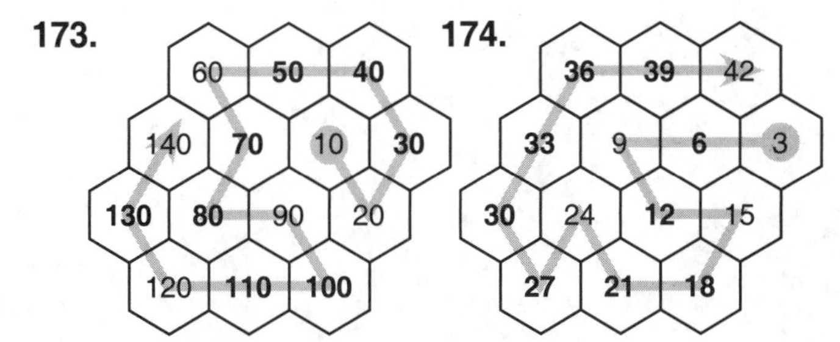

175. **176.**

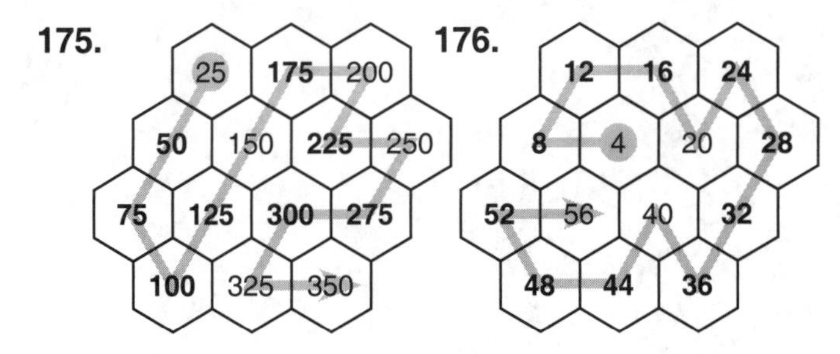

177. **178.**

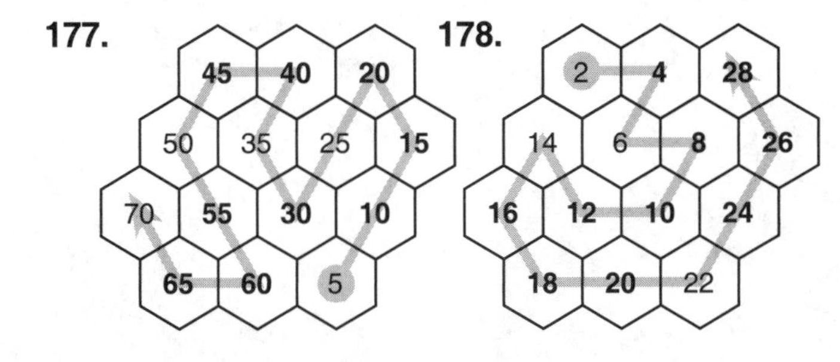

179. **180.**

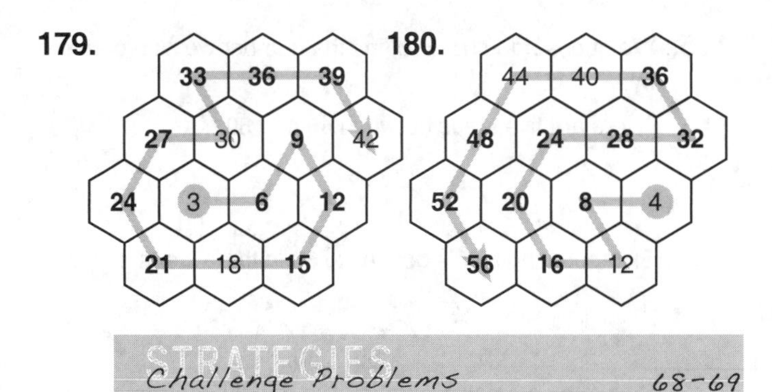

STRATEGIES
Challenge Problems 68-69

181. $(777+444)$ is 444 *more* than 777.
$(777-444)$ is 444 *less* than 777.

So, the difference between $(777+444)$ and $(777-444)$ is $444+444=$ **888**.

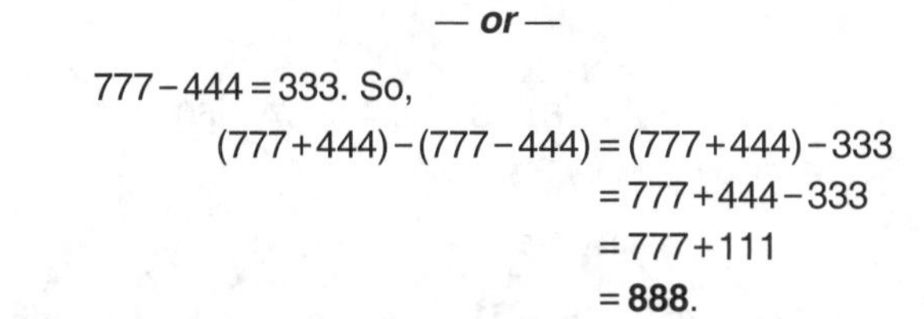

— *or* —

$777-444=333$. So,

$$\begin{aligned}(777+444)-(777-444) &= (777+444)-333\\ &= 777+444-333\\ &= 777+111\\ &= \mathbf{888}.\end{aligned}$$

182. Each day, Charlie has $36-31=5$ more berries than the day before. To find how many berries Charlie has after 7 days, we skip-count to add seven 5's.

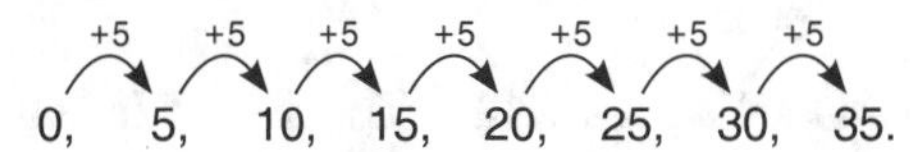

So, after 7 days, Charlie has **35** berries in his bucket.

183. 200 is 2 more than 198. So, we need to fill the circles so that we add 2 more than we take away.

If we take away 55 then add 56, we add 1 more than we take away.

If we take away 57 then add 58, we add 1 more than we take away.

So, if we fill the circles as shown below, we add $1+1=2$ more than we take away, giving a true equation.

$198 \ominus 55 \oplus 56 \ominus 57 \oplus 58 = 200.$

This is the only possible solution.

184. There are 11 groups of dots in the diagram. Each group has $7-4=3$ more black dots than white dots.

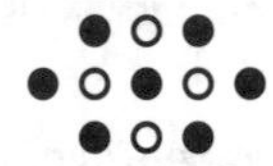

So, the diagram has $3+3+3+3+3+3+3+3+3+3+3$ more black dots than white dots. We skip-count to add eleven 3's:

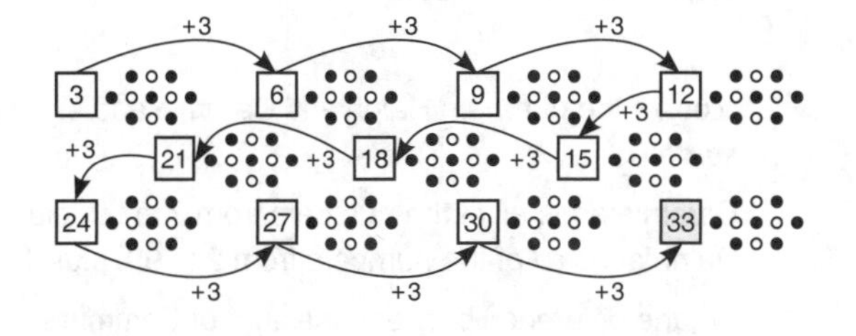

So, there are **33** more black dots than white dots.

185. We guess-and-check to find the number we skip-count by.

There are five steps between 45 and 60.

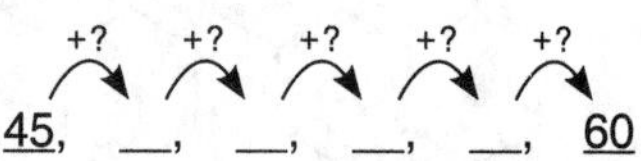

Skip-counting by 5's gets us to 60 in three steps, not five.

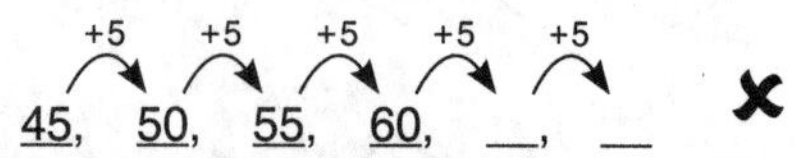

So, we try a smaller number. We cannot get from 45 to 60 skip-counting by 4's.

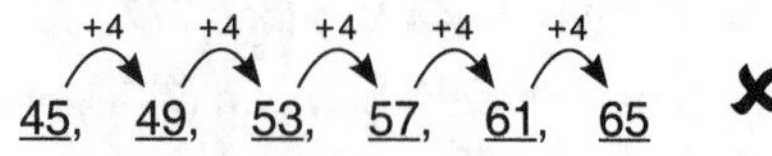

But, skip-counting by 3's works!

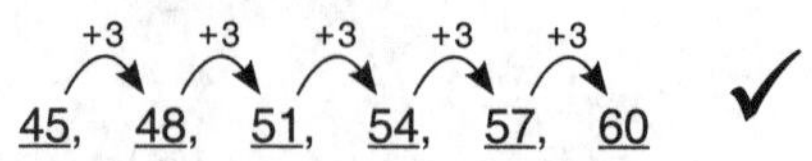

So, we skip-count by 3's.

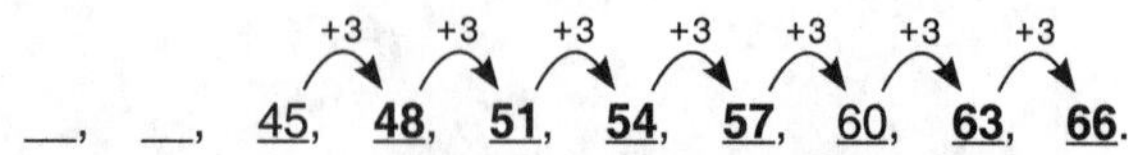

Since the number to the right of 45 is 3 *more* than 45, the number to the left of 45 is 3 *less* than 45. We continue this pattern to the left.

−3 −3

39, **42**, 45, **48**, **51**, **54**, **57**, 60, **63**, **66**.

186. Adding a number then taking away that number is the same as doing nothing. Since the same number fills each blank, the first two blanks in the expression cancel. This simplifies the equation to 44+____=100.

Since 44+56=100, we fill the blanks with 56.

44+**56**−**56**+**56**=100.

187. There are 100 numbers from 1 to 100, so there are 99 numbers from 2 to 100.

So, Ralph and Cammie each add 99 numbers.

Cammie's first number (2) is 1 more than Ralph's first number (1).

Cammie's second number (3) is 1 more than Ralph's second number (2).

Cammie's third number (4) is 1 more than Ralph's third number (3).

If we list their numbers in order, each of Cammie's 99 numbers is 1 more than one of Ralph's 99 numbers.

So, the sum of Cammie's numbers is **99** greater than the sum of Ralph's numbers.

— or —

Most of the numbers Ralph and Cammie add are the same.

Cammie adds all of the numbers from 2 to 99, plus 100. Ralph adds all of the numbers from 2 to 99, plus 1.

So, the difference between the sum of Cammie's numbers and the sum of Ralph's numbers is 100−1=**99**.

188. We start by subtracting (20+19)−(19+18).

20+19 is 2 more than 19+18. So, (20+19)−(19+18)=2.

Similarly, (18+17)−(17+16)=2 and (16+15)−(15+14)=2.

So, we have

$$\underbrace{(20+19)-(19+18)}_{2}+\underbrace{(18+17)-(17+16)}_{2}+\underbrace{(16+15)-(15+14)}_{2}$$

= 2 + 2 + 2

= **6.**

Chapter 9: Solutions

Basics 71

1. Even numbers are the numbers that end in 0, 2, 4, 6, or 8. We circle the even numbers as shown below.

23  26  30 41 55  78

2. Odd numbers are the numbers that end in 1, 3, 5, 7, or 9. We circle the odd numbers as shown below.

229 442  531 577 790  983

3. The largest two-digit number is 99. Since 99 ends in 9, it is odd. So, the largest two-digit *odd* number is **99**.

4. The largest three-digit number is 999. Since 999 ends in 9, it is odd. The next-largest three-digit number is 998. Since 998 ends in 8, it is even.

So, the largest three-digit *even* number is **998**.

5. We list the odd numbers between 46 and 64:

47, 49, 51, 53, 55, 57, 59, 61, 63.

All together, there are **9** odd numbers between 46 and 64.

Twos 72–73

6. If we can split the dots into two equal groups, then the number of dots is even.

If we try to split the dots into two equal groups and there is one dot left over, then the number of dots is odd.

We use this reasoning to circle the figures with an even number of dots, as shown below.

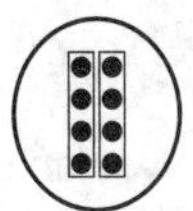 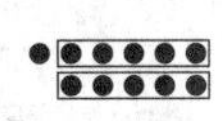 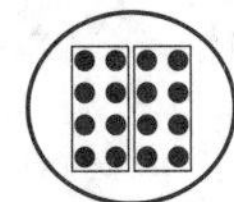 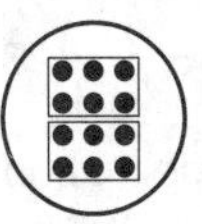 

7. If we can split the dots into two equal groups *or* into groups of two, then the number of dots is even.

If we try to split the dots into two equal groups *or* into groups of two and there is one dot left over, then the number of dots is odd.

We use this reasoning to circle the figures with an odd number of dots, as shown below.

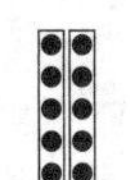 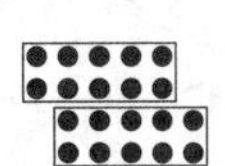 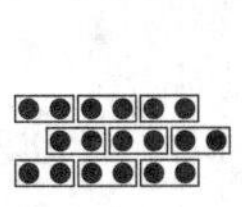 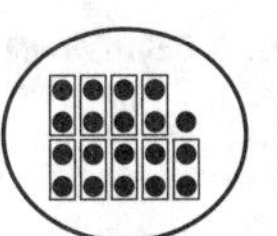 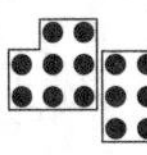

8. We use the strategies discussed in the previous two problems to circle the figures with an even number of dots.

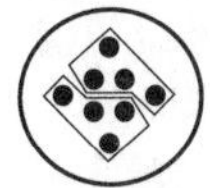  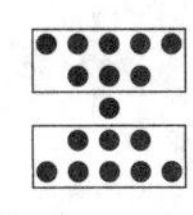 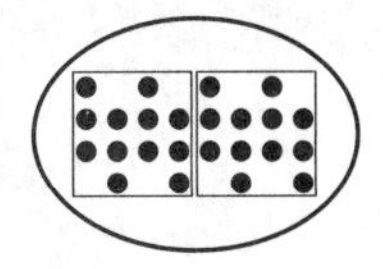 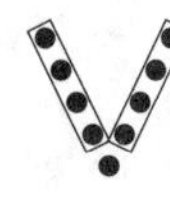

9. If a number can be split into groups of two, then it is even. Since every jackalope has two horns, the total number of horns is **even**.

10. Every monster invited brings one friend. So, the monsters come in groups of two. This means that the number of guests at the party is even.

When we include Mergle, we get one extra monster. One more than an even number is an odd number. So, the total number of monsters at Mergle's party is **odd**.

11. Each card has two shapes: one on the front and one on the back. Since the shapes come in groups of two, the total number of shapes is **even**.

12. Each team has the same number of players. So, when two teams play each other, the total number of players can be split into two equal groups: one group for each team. This means that the total number of players is even.

When we include the referee, we get one extra beast. One more than an even number is an odd number. So, the total number of beasts on the field is **odd**.

13. Every high five takes place between two monsters. So, every high five will be counted twice: once for each of the two monsters.

For example, if Alex and Grogg high-five, that high five will be counted once by Alex and once by Grogg.

Since every high five gets counted twice, the total number of high fives counted will be **even**.

Half 74–75

14. We can split 26 into two equal groups of 13.

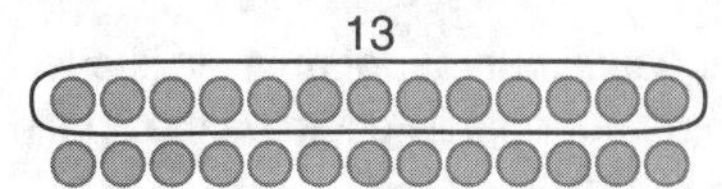

So, half of 26 is **13**.

— or —

Since $13+13=26$, half of 26 is **13**.

15. Since $24+24=48$, half of 48 is **24**.

16. Since $203+203=406$, half of 406 is **203**.

17. Since $134+134=268$, half of 268 is **134**.

18. Since $17+17=34$, half of 34 is **17**.

— or —

34 is $30+4$. Half of 30 is 15, and half of 4 is 2.

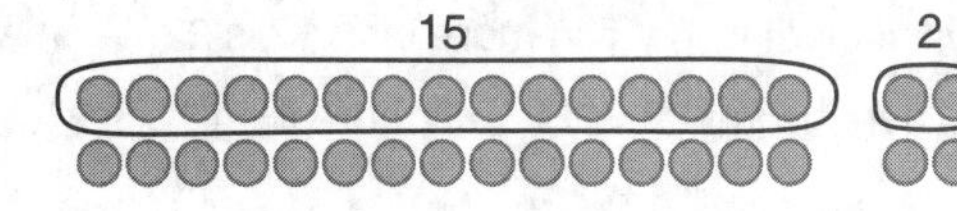

So, half of 34 is $15+2=$ **17**.

19. Since 26+26=52, half of 52 is **26**.

— or —

52 is 50+2. Half of 50 is 25, and half of 2 is 1.

So, half of 52 is 25+1=**26**.

20. Since 125+125=250, half of 250 is **125**.

— or —

250 is 200+50. Half of 200 is 100, and half of 50 is 25.

So, half of 250 is 100+25=**125**.

21. 770 is 700+70. Half of 700 is 350, and half of 70 is 35.

So, half of 770 is 350+35=**385**.

22. When we double a number, then find half of the result, we get the same number that we started with.

For example, doubling 2 gives 4, and half of 4 is 2. Doubling 150 gives 300, and half of 300 is 150.

So, doubling 479, then finding half of the result gives **479**.

23. Half of 144 is 72. Half of 72 is 36. Half of 36 is 18.

So, half of half of half of 144 is **18**.

24. When Vike and Ash share the marbles equally, each monster will get half of the total number of marbles.

There are 147+93=240 total marbles. Since 120+120=240, half of 240 is 120. So, Vike and Ash will each get **120** marbles.

25. The 48 boys who wear glasses are half of the total number of boys. So, there are 48+48=96 boys.

The 96 boys are half of the total number of students. So, there are 96+96=**192** total students.

26. Exactly half of the animals on the farm are chickens. So, the total number of animals can be split into two equal groups: one group of chickens, and one group of all the other animals.

Any even number can be split into two equal groups. No odd number can be split into two equal groups.

So, the total number of animals is even. Each circled even number below could be the total number of animals.

(18) 25 (76) (98) 243 777 (952)

27. We begin by filling in the blanks to the right of 36.

The number to the right of 36 is 10 more than half of 36. Half of 36 is 18. Ten more than 18 is 28.

____, ____, 52, 36, **28**, ____, 22, ____

Then, half of 28 is 14. Ten more than 14 is 24.

____, ____, 52, 36, **28**, **24**, 22, ____

We check that the next number is 22:
Half of 24 is 12. Ten more than 12 is 22. ✓

Then, half of 22 is 11. Ten more than 11 is 21.

____, ____, 52, 36, **28**, **24**, 22, **21**

Next, we work backwards to fill the blanks to the left of 52.

52 is 10 more than half the number to its left.
52 is 10 more than 42. So, 42 is half the number to the left of 52. Since 42+42=84, the number left of 52 is 84.

____, **84**, 52, 36, **28**, **24**, 22, **21**

84 is 10 more than half the number to its left.
84 is 10 more than 74. So, 74 is half the number to the left of 84. Since 74+74=148, the number left of 84 is 148.

148, **84**, 52, 36, **28**, **24**, 22, **21**

ODDS & EVENS

Addition 76-79

28. We compute each sum as shown below:

8+13=21.
8+21=29.
8+24=32.
8+42=50.
8+53=61.

Only 32 and 50 are even. So, we circle the sums with even results as shown.

8+13 8+21 (8+24) (8+42) 8+53

Notice that adding an even number to 8 gives an even result. Adding an odd number to 8 gives an odd result.

29. We compute each sum as shown below:

23+6=29.
23+12=35.
23+34=57.
23+53=76.
23+89=112.

Only 76 and 112 are even. So, we circle the sums with even results as shown.

23+6 23+12 23+34 (23+53) 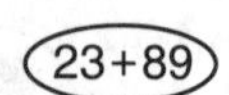

Notice that adding an even number to 23 gives an odd result. Adding an odd number to 23 gives an even result.

30. We compute each sum as shown below:

14+11=25.
21+11=32.
28+11=39.
33+11=44.
45+11=56.

Only 25 and 39 are odd. So, we circle the sums with odd results as shown.

(14+11) 21+11 (28+11) 33+11 45+11

Notice that adding 11 to an even number gives an odd result. Adding 11 to an odd number gives an even result.

31. We compute each sum as shown below:

$19+14=33$.
$32+14=46$.
$38+14=52$.
$55+14=69$.
$91+14=105$.

Only 33, 69, and 105 are odd. So, we circle the sums with odd results as shown.

19+14 32+14 38+14 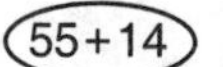

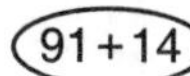

Notice that adding 14 to an even number gives an even result. Adding 14 to an odd number gives an odd result.

32. Every even number can be split into groups of two. So, when we add two even numbers, we always get groups of two.

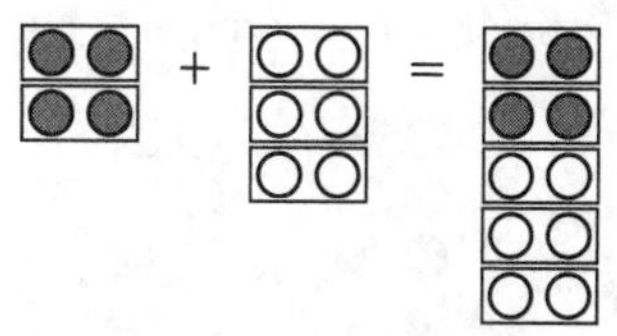

So, an even plus an even is always **even**.

33. Every odd number can be split into groups of two with one extra. So, when we add two odd numbers, we get groups of two with two extras. These two extras can be combined into another group of two!

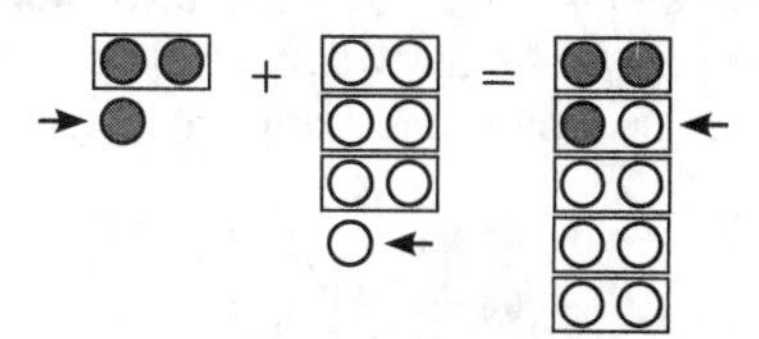

So, an odd plus an odd is always **even**.

34. Every even number can be split into groups of two. Every odd number can be split into groups of two with one extra.

So, when we add an even number plus an odd number, we get groups of two with one extra.

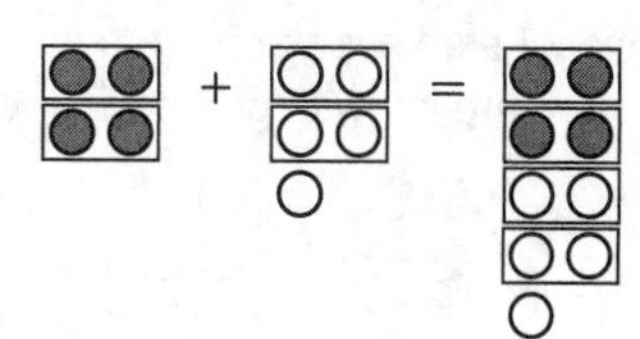

So, an even plus an odd is always **odd**.

35. Every odd number can be split into groups of two with one extra. Every even number can be split into groups of two.

So, when we add an odd number plus an even number, we get groups of two with one extra.

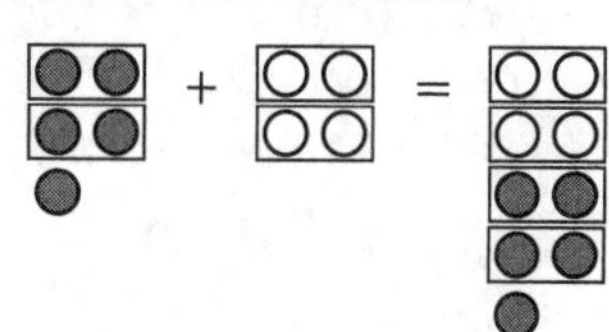

So, an odd plus an even is always **odd**.

36. In Problems 32-35, we learned the facts below.

even + even = even
odd + odd = even
even + odd = odd
odd + even = odd

We use these facts to determine whether each sum is even or odd, as shown.

$419+853=$ odd+odd = even.
$824+786=$ even+even = even.
$742+217=$ even+odd = odd.
$819+777=$ odd+odd = even.
$203+698=$ odd+even = odd.

We circle the sums with even results.

(419+853) (824+786) 742+217 (819+777) 203+698

37. We consider each sum as shown below.

$348+672=$ even+even = even.
$709+583=$ odd+odd = even.
$722+313=$ even+odd = odd.
$639+920=$ odd+even = odd.
$253+129=$ odd+odd = even.

We circle the sums with odd results.

348+672 709+583 (722+313) (639+920) 253+129

38. 244 more than an even number is the sum of two evens, which is even. So, the numbers we circle must be even.

Every even number greater than 244 is 244 more than an even number. So, we circle 408 and 992.

(408) 793 737 387 (992)

We could subtract 244 from the circled answers to find which even numbers they are 244 more than.

39. 159 more than an odd number is the sum of two odds, which is even. So, the numbers we circle must be even.

Every even number greater than 159 is 159 more than an odd number. So, we circle 284 and 646.

331 (284) 931 (646) 445

We could subtract 159 from the circled answers to find which odd numbers they are 159 more than.

40. We can only get an odd sum by adding an even and an odd. 4, 6, and 8 are even, and 7 is odd. So, 7 must be placed in the center triangle that shares a side with each of the other three triangles.

It does not matter how we place 4, 6, and 8 in the other three triangles. We show all possible arrangements below.

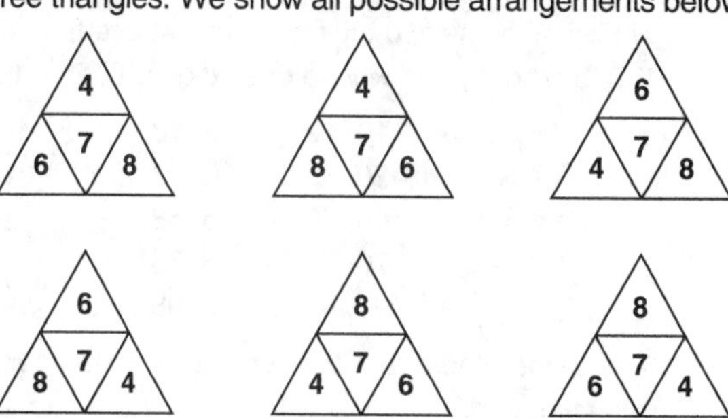

41. Five years ago, Finn's age was odd. So, his age today is some odd number plus 5. The sum of two odds is even, so Finn's age today is **even**.

42. The two numbers on either side of Abe's page are consecutive. So, there are two possibilities:

- The first page number is even and the next is odd.
- The first page number is odd and the next is even.

So, there is one even page number and one odd page number. The sum of an even and an odd is odd. So, the sum of the two page numbers is **odd**.

43. Grogg and Lizzie have the same number of pencils. Since the sum of two equal groups is even, the total number of pencils is even.

The only two even numbers are 34 and 56. So, Grogg and Lizzie could have a total of 34 pencils ($17+17=34$) or 56 pencils ($28+28=56$).

25 17 (34) (56) 41 97

44. If Alex and Winnie had the same number of pencils, then the total number of pencils would be even. Since Alex has 5 more pencils than Winnie, the total number of pencils is some even number plus 5.

The sum of an even and an odd is odd. So, the total number of pencils must be odd. The only odd numbers are 55, 13, 29, and 81.

So, Alex and Winnie could have a total of 55 pencils ($30+25=55$), 13 pencils ($9+4=13$), 29 pencils ($17+12=29$), or 81 pencils ($43+38=81$).

40 (55) 24 (13) (29) (81)

— or —

Alex has 5 more pencils than Winnie. So, if Winnie has an even number of pencils, Alex has an odd number of pencils. If Winnie has an odd number of pencils, Alex has an even number of pencils.

So, the total number of pencils is the sum of an even and an odd, which is odd. 55, 13, 29, and 81 are the only odd numbers.

So, Alex and Winnie could have a total of 55 pencils ($30+25=55$), 13 pencils ($9+4=13$), 29 pencils ($17+12=29$), or 81 pencils ($43+38=81$).

40 (55) 24 (13) (29) (81)

45. Since we are considering *even* two-digit numbers, the ones digit must be even: 0, 2, 4, 6, or 8. We can only get an odd digit-sum if the tens digit is odd: 1, 3, 5, 7, or 9.

We list all of the two-digit numbers whose tens digit is 1, 3, 5, 7, or 9, and whose ones digit is 0, 2, 4, 6, or 8.

10, 12, 14, 16, 18,
30, 32, 34, 36, 38,
50, 52, 54, 56, 58,
70, 72, 74, 76, 78,
90, 92, 94, 96, 98.

All together, there are **25** even two-digit numbers with an odd digit-sum.

ODDS & EVENS

Subtraction 80-81

46. Every even number can be split into groups of two. So, when we subtract two even numbers, we take some groups of two away from another number of groups of two. This leaves us with fewer groups of two.

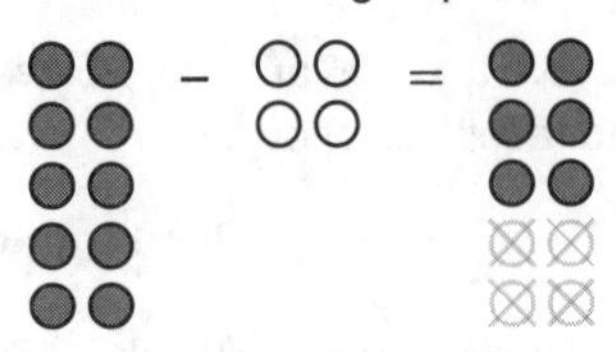

So, an even minus an even is always **even**.

47. Every odd number can be split into groups of two with one extra. When we take away an odd number from an odd number, we take away some groups of two and one extra. This only leaves groups of two.

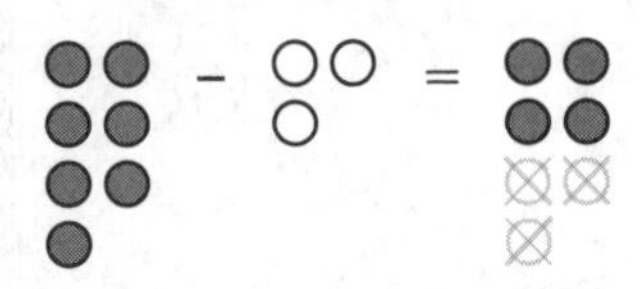

So, an odd minus an odd is always **even**.

48. Every even number can be split into groups of two. Every odd number can be split into groups of two with one extra.

When we take an odd number away from an even number, we take away some groups of two and one extra. This leaves one extra that is not in a group of two.

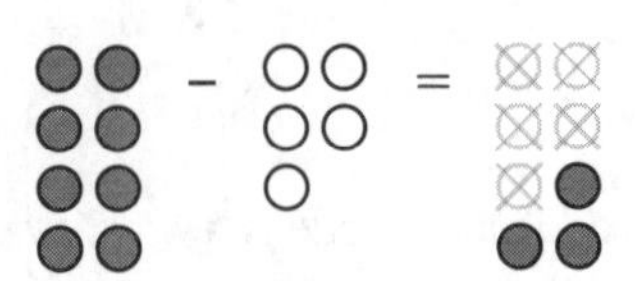

So, an even minus an odd is always **odd**.

49. Every odd number can be split into groups of two with one extra. Every even number can be split into groups of two.

When we take an even number away from an odd number, we are left with groups of two and one extra.

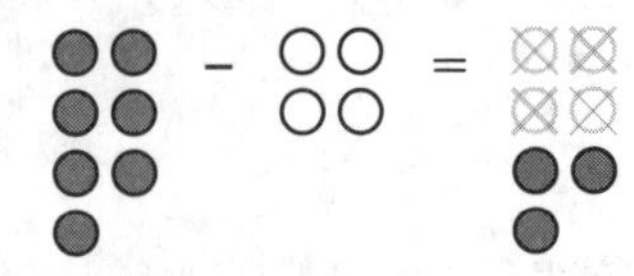

So, an odd minus an even is always **odd**.

The patterns for subtracting evens and odds are the same as the patterns for adding evens and odds!

even + even = even	*even − even = even*
odd + odd = even	*odd − odd = even*
even + odd = odd	*even − odd = odd*
odd + even = odd	*odd − even = odd*

50. We consider each difference as shown below.

719 − 183 = odd − odd = even.
454 − 176 = even − even = even.
298 − 219 = even − odd = odd.
459 − 273 = odd − odd = even.
833 − 198 = odd − even = odd.

We circle the differences with odd results.

719 − 183 454 − 176 (298 − 219) 459 − 273 (833 − 198)

51. We consider each difference as shown below.

848 − 672 = even − even = even.
939 − 560 = odd − even = odd.
709 − 583 = odd − odd = even.
722 − 313 = even − odd = odd.
253 − 129 = odd − odd = even.

We circle the differences with even results.

(848 − 672) 939 − 560 (709 − 583) 722 − 313 (253 − 129)

52. It is possible for both results to be odd. For example, if the numbers are 3 and 4, Grogg's result is 3 + 4 = 7 and Winnie's result is 4 − 3 = 1. Both 7 and 1 are odd.

It is possible for both results to be even. For example, if the numbers are 3 and 5, Grogg's result is 3 + 5 = 8 and Winnie's result is 5 − 3 = 2. Both 8 and 2 are even.

It is impossible for one result to be odd and the other result to be even. This is because the rules for subtracting evens and odds are the same as the rules for adding evens and odds. If the sum of two numbers is odd, the difference is also odd. Similarly, if the sum is even, the difference is also even.

We circle the impossible statement as shown.

Both results are odd. Both results are even. (One result is odd, one result is even.)

53. Since both boxes hold the same number of cookies, the total number of cookies is even.

Marcus eats 17 cookies. So, the number of cookies left is some even number minus 17. An even minus an odd is odd. So, the number of cookies left is **odd**.

54. We can only get an odd difference if one number is even and the other number is odd. 234, 456, and 678 are even, and 567 is odd. So, 567 must be placed in the center triangle that shares a side with each of the other three triangles.

It does not matter how we place 234, 456, and 678 in the other triangles. We show all possible arrangements below.

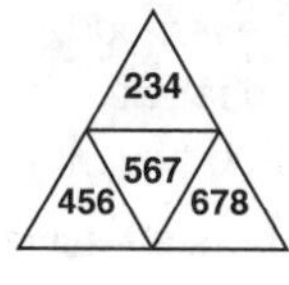

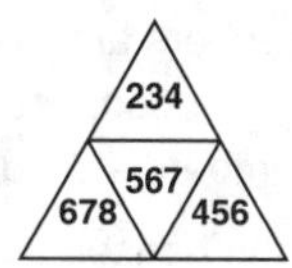

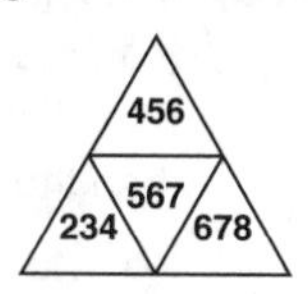

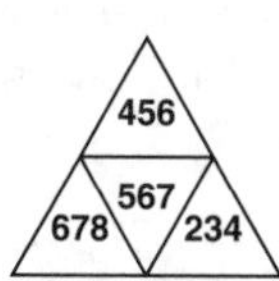

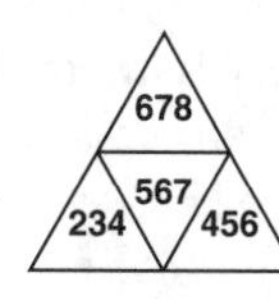

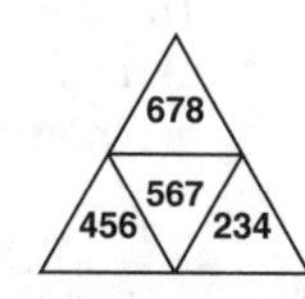

55. Adding from left to right, we have

123 + 234 + 345
= odd + even + odd
= odd + odd
= even.

So, 123 + 234 + 345 is **even**.

— or —

We can add numbers in any order. So, we have

123 + 234 + 345
= odd + even + odd
= even + even
= even.

So, 123 + 234 + 345 is **even**.

56. Adding from left to right, we have

172 + 283 + 394
= even + odd + even
= odd + even
= odd.

So, 172 + 283 + 394 is **odd**.

— or —

We can add numbers in any order. So, we have

172 + 283 + 394
= even + odd + even
= even + odd
= odd.

So, 172 + 283 + 394 is **odd**.

57. We can add numbers in any order. So, we have

22 + 44 + 66 + 88
= even + even + even + even
= even + even
= even.

So, 22 + 44 + 66 + 88 is **even**.

58. We can add numbers in any order. So, we have

19 + 29 + 39 + 49 + 59
= odd + odd + odd + odd + odd
= even + even + odd
= even + odd
= odd.

So, 19 + 29 + 39 + 49 + 59 is **odd**.

59. We can add numbers in any order. So, we have

12 + 34 + 56 + 78 + 90
= even + even + even + even + even
= even + even + even
= even + even
= even.

So, 12+34+56+78+90 is **even**.

60. We can add numbers in any order. So, we have

43 + 18 + 42 + 94 + 78
= odd + even + even + even + even
= odd + even + even
= odd + even
= odd.

So, 43+18+42+94+78 is **odd**.

61. An even plus an even is even.
So, the sum of 2 evens is **even**.

The sum of 2 evens is even. Adding a 3rd even gives an even plus an even, which is even.
So, the sum of 3 evens is **even**.

The sum of 3 evens is even. Adding a 4th even gives an even plus an even, which is even.
So, the sum of 4 evens is **even**.

The sum of 4 evens is even. Adding a 5th even gives an even plus an even, which is even.
So, the sum of 5 evens is **even**.

We see that the sum of even numbers is always even, no matter how many even numbers there are.

So, the sum of 100 evens is **even**,
and the sum of 333 evens is **even**.

62. An odd plus an odd is even.
So, the sum of 2 odds is **even**.

The sum of 2 odds is even. Adding a 3rd odd gives an even plus an odd, which is odd.
So, the sum of 3 odds is **odd**.

The sum of 3 odds is odd. Adding a 4th odd gives an odd plus an odd, which is even.
So, the sum of 4 odds is **even**.

The sum of 4 odds is even. Adding a 5th odd gives an even plus an odd, which is odd.
So, the sum of 5 odds is **odd**.

We see that the sum of odd numbers alternates between even and odd. If there is an even number of odds, then the sum is even. If there is an odd number of odds, then the sum is odd.

Since 99 is odd, the sum of 99 odds is **odd**.
Since 250 is even, the sum of 250 odds is **even**.

63. There is an odd number of pages, and each page contains an odd number of stickers. So, the total number of stickers is the sum of an odd number of odds.

In the previous problem, we learned that the sum of an odd number of odds is always odd.

So, the total number of stickers in Yuri's book is **odd**.

64. All but one bus have an even number of wheels. In Problem 61, we learned that the sum of even numbers is always even. So, the total number of wheels on the *even*-wheeled buses is even.

When we add the wheels from the *odd*-wheeled bus, we get an even plus an odd, which is odd.

So, the total number of bus wheels is **odd**.

65. There is an even number of hands, and every hand has an odd number of fingers. So, the total number of fingers is the sum of an even number of odds.

In Problem 62, we learned that the sum of an even number of odds is always even.

So, the total number of fingers is **even**.

66. When Sam adds an odd number of odds, he gets an odd result. He subtracts from this an even number of evens, which is even.

So, Sam's final result is an odd minus an even, which is **odd**.

67. For the sum to be odd, we must add an odd number of odds. So, either 1 or 3 of the four numbers must be odd.

To get the largest sum with 1 odd, we add the largest odd and the three largest evens: 35+50+20+12 = 117.

To get the largest sum with 3 odds, we add the three largest odds and the largest even: 35+25+5+50 = 115.

So, **117** is the largest odd sum we can get by adding four of the given numbers.

68. If the first of the three consecutive numbers is even, then the three numbers are even-odd-even.
If the first of the three consecutive numbers is odd, then the three numbers are odd-even-odd.

To get an odd sum, there must be an odd number of odds. So, any three consecutive numbers that are even-odd-even will give an odd sum. **2+3+4** = 9 is one example. There are many others!

69. To get an even sum, there must be an even number of odds. So, any three consecutive numbers that are odd-even-odd will give an even sum. **1+2+3** = 6 is one example. There are many others!

70. If the first of the four consecutive numbers is even, then the four numbers are even-odd-even-odd.
If the first of the four consecutive numbers is odd, then the four numbers are odd-even-odd-even.

So, two of the four consecutive numbers are always odd. This means that the sum of four consecutive numbers is always even. So, it is **impossible** for four consecutive numbers to have an odd sum.

71. In the previous problem, we learned that for any four consecutive numbers, two will be odd and two will be even. So, the sum of four consecutive numbers is always even.

1+**2**+**3**+**4** = 10 and **2**+**3**+**4**+**5** = 14 are two examples. There are many others!

72. The largest possible sum of three different 2-digit numbers is 99+98+97. But, since exactly two of these numbers are odd, this sum is even.

The second-largest possible sum of three different 2-digit numbers is 99+98+96. Since exactly one of these numbers is odd, this sum is odd.

So, 99+98+96 = **293** is the largest odd result we can get by adding three different 2-digit numbers.

73. Four of the seven numbers are odd. For each group to have an even sum, there must be an even number of odds in each group. This gives two possibilities:

- One group has 0 odds and the other has 4 odds.
- One group has 2 odds and the other has 2 odds.

Since the first and last numbers are each odd, we cannot draw a line so that one group has 0 odds. So, we draw a line so that each group has 2 odds. There is only one way to do this, as shown below.

311	340	156	208	253 |	101	359
odd	even	even	even	odd	odd	odd

74. We are given that 15 = 5+4+3+2+1.

Changing the +1 to −1 gives 5+4+3+2−1 = 13.
Changing the +2 to −2 gives 5+4+3−2+1 = 11.
Changing the +3 to −3 gives 5+4−3+2+1 = 9.
Changing the +4 to −4 gives 5−4+3+2+1 = 7.

We notice that any time we change a + to a −, the result decreases by an even amount. For example, changing +1 to −1 decreases the result by 2, changing +2 to −2 decreases the result by 4, and so on.

We start with a sum of 15 using all + signs. Decreasing 15 by an even amount always gives an odd result. So, it is impossible to get an even result!

We use the strategies described above to fill the blanks as shown.

15 = 5+4+3+2+1	14 = **Impossible**
13 = **5+4+3+2−1**	12 = **Impossible**
11 = **5+4+3−2+1**	10 = **Impossible**
9 = **5+4−3+2+1**	8 = **Impossible**
7 = 5−4+3+2+1	6 = **Impossible**
5 = **5−4+3+2−1**	4 = **Impossible**
3 = **5−4+3−2+1**	2 = **Impossible**
1 = **5−4+3−2−1**	0 = **Impossible**

You may have gotten some of the odd results above using different expressions.

ODDS & EVENS

8's and 9's 86-87

In each solution below, we show one set of steps to solve the problem. You may have used a different set of steps to arrive at the same final answer.

75. We can tell whether a sum is even or odd by how many odd numbers are in the sum. If there is an even number of odds, the sum will be even. If there is an odd number of odds, the sum will be odd.

Since 9 is odd and 8 is even, there must be an odd number of 9's in each row to get an odd sum. It does not matter how many 8's are in each row.

We fill the blanks so that each row has an odd number of 9's as shown below.

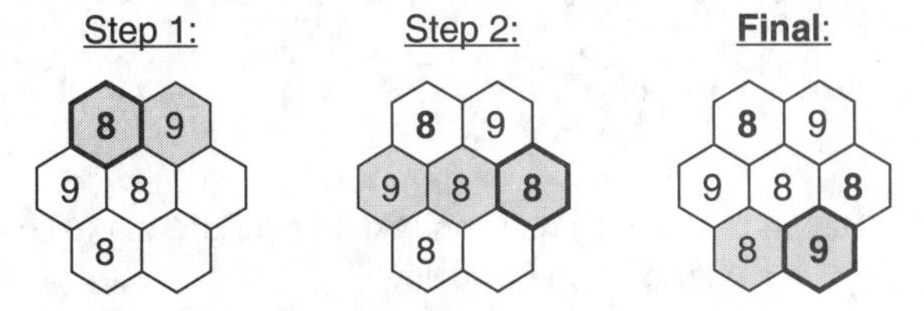

76. Each row must have an odd number of 9's. We complete the puzzle as shown below.

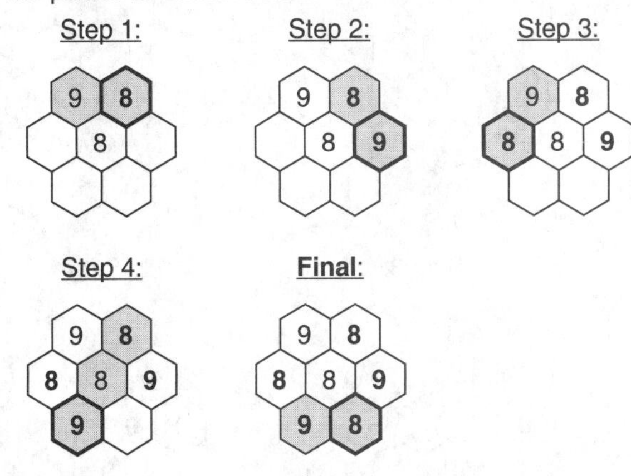

77. Each row must have an odd number of 9's. We complete the puzzle as shown below.

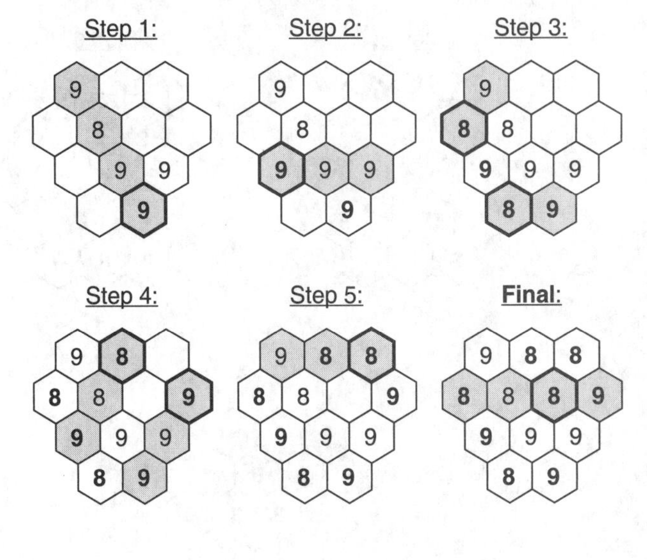

78. Each row must have an odd number of 9's. We complete the puzzle as shown below.

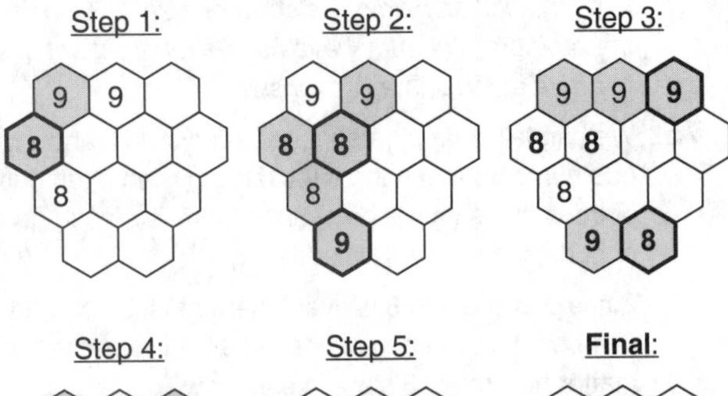

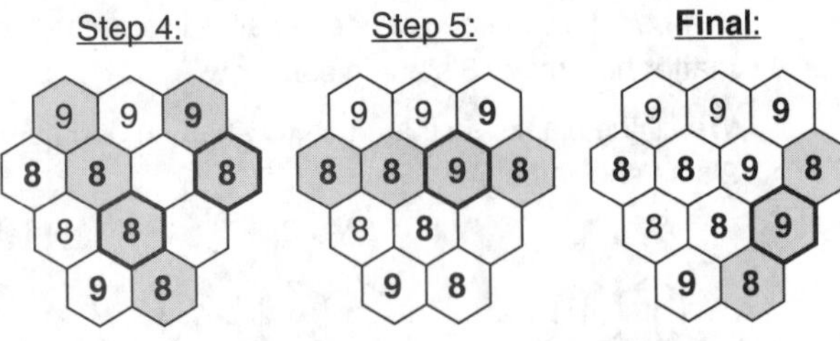

79. Each row must have an odd number of 9's. We complete the puzzle as shown below.

Step 1:

Step 2:

Step 3:

Step 4:

Step 5:

Step 6:

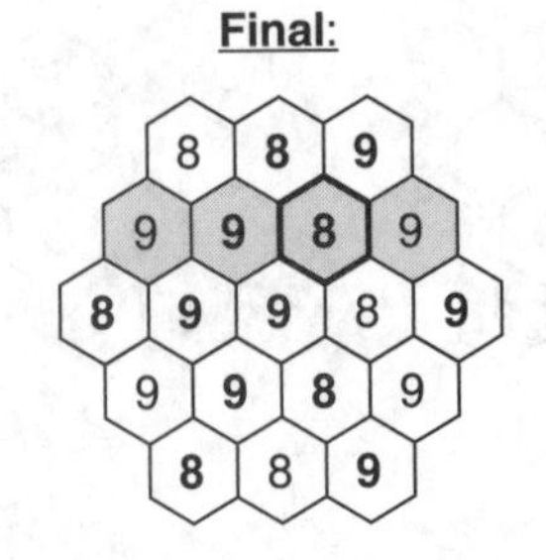

We use the strategies from the previous solutions to complete the puzzles below.

80. **81.**

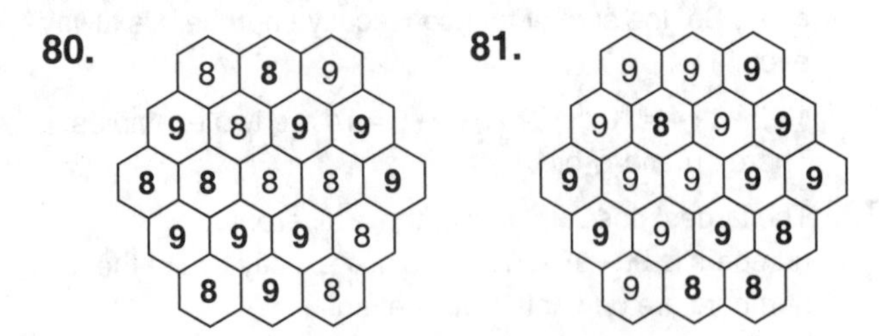

82. We fill the first few hexagons as shown.

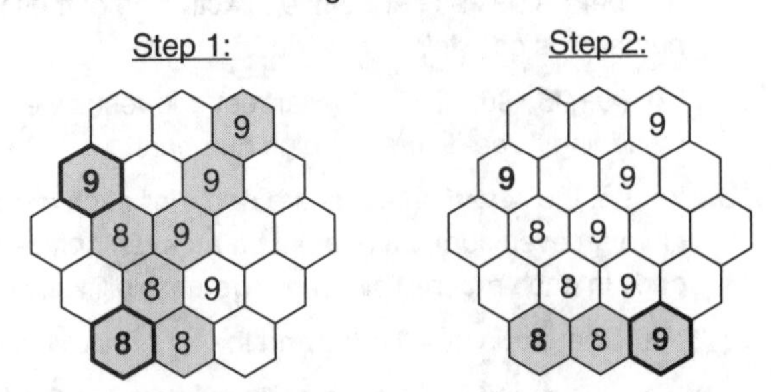

Then, we consider the leftmost hexagon of the middle row. If this hexagon is 8, then we can fill the remaining hexagons with the following steps.

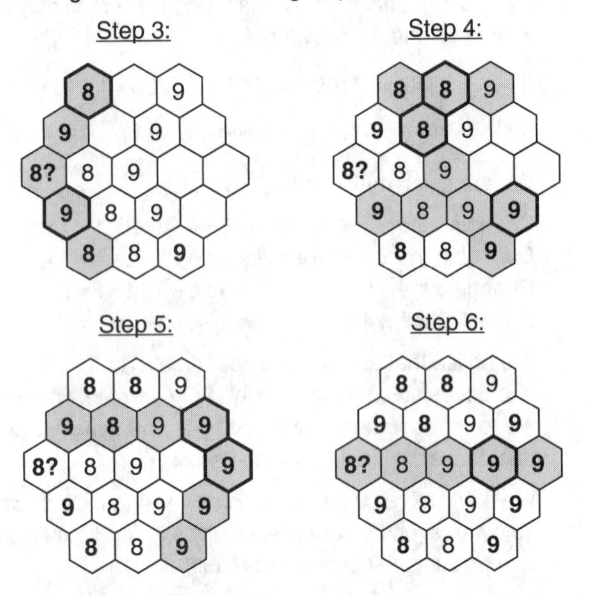

All rows in all directions have an odd sum. So, this works!

If we fill the leftmost hexagon of the middle row with 9, then try to fill the remaining hexagons, we eventually get a row with an even sum.

So, the only possible answer is shown below.

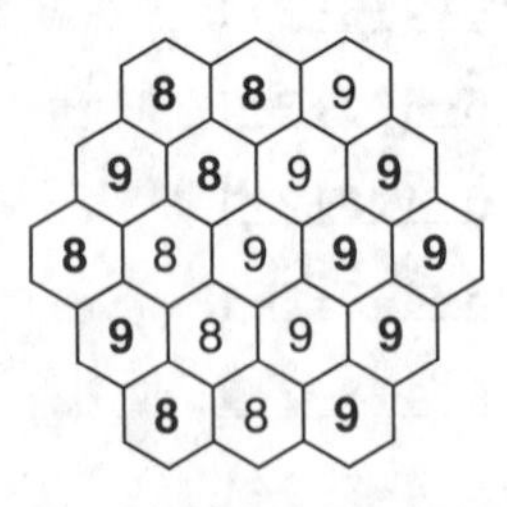

If you did not get the answer above, double-check your rows to find any even sums!

ODDS & EVENS — Back and Forth (88–89)

83. We list the person who feeds Spark for the first few days of March and look for a pattern.

On March 1st, Ell feeds Spark.
On March 2nd, Mel feeds Spark.
On March 3rd, Ell feeds Spark.
On March 4th, Mel feeds Spark.
On March 5th, Ell feeds Spark.
On March 6th, Mel feeds Spark.

We notice that Ell feeds Spark on the odd-numbered days, and Mel feeds Spark on the even-numbered days.

Since 31 is odd, **Ell** feeds Spark on March 31st.

84. The 1st bead on Ruth's string is red.
The 2nd bead on Ruth's string is blue.
The 3rd bead on Ruth's string is red.
The 4th bead on Ruth's string is blue.

This pattern continues.
All of the odd-numbered beads are red.
All of the even-numbered beads are blue.

Since 50 is even, the 50th bead on Ruth's string is **blue**.

85. Alex takes his 1st step with his left foot.
Alex takes his 2nd step with his right foot.
Alex takes his 3rd step with his left foot.
Alex takes his 4th step with his right foot.

Alex takes his odd-numbered steps with his left foot and his even-numbered steps with his right foot. Since 250 is even, Alex takes his 250th step with his **right** foot.

86. After 1 stream crossing, Beast Grylls ends up on the *opposite* side of the river from camp.

After 2 stream crossings, Beast Grylls ends up on the *same* side of the river from camp.

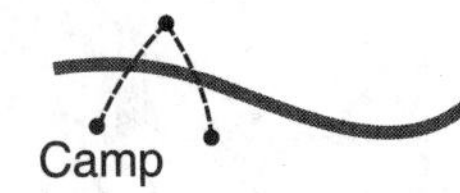

After 3 stream crossings, Beast Grylls ends up on the *opposite* side of the river from camp.

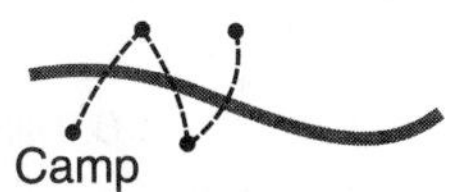

After 4 stream crossings, Beast Grylls ends up on the *same* side of the river from camp.

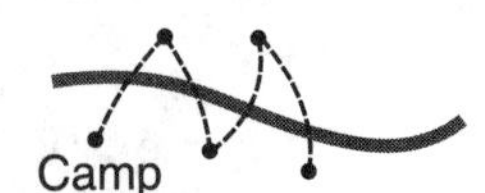

If Beast Grylls crosses an odd number of times, he will be on the opposite side of the river from camp. So, Beast Grylls will be on the *opposite* side after 27 crossings. So, **yes**, he will need to cross again to get to camp.

87. The 1st number Grogg says is 30, which is even.
The 2nd number Grogg says is 33, which is odd.
The 3rd number Grogg says is 36, which is even.
The 4th number Grogg says is 39, which is odd.
The 5th number Grogg says is 42, which is even.

Grogg's numbers alternate between even and odd.

The "oddth" numbers Grogg says are even: the 1st, 3rd, 5th, and so on. Since 75 is odd, the 75th number Grogg says is **even**.

88. The Crocadillos were in the lead after ten minutes.

After the 1st lead change, the Elephrogs were in the lead.
After the 2nd lead change, the Crocadillos were in the lead.
After the 3rd lead change, the Elephrogs were in the lead.
After the 4th lead change, the Crocadillos were in the lead.

If the number of lead changes is odd, the Elephrogs are in the lead. After the 1st lead change, there were 15 more lead changes. That makes 16 lead changes all together. Since 16 is even, the Crocadillos had the final lead. So, the **Crocadillos** won.

89. Since the little monsters alternate boy-girl-boy-girl, we can group them into pairs, with one boy and one girl in each pair.

175 is odd. So, there will be one monster left over after making our pairs. Since the first monster in line is a boy, the first monster in every pair is a boy. So, the monster left over after making every pair will be a boy.

There is an equal number of boy and girl monsters in all of our pairs. So, the one extra boy monster gives more boys than girls in line all together.

There are more girls than boys in line.

(There are more boys than girls in line.)

There is the same number of girls as boys in line.

90. Starting from the outside of the shape, we can trace a path to reach the top-left star. So, this star is outside of the shape.

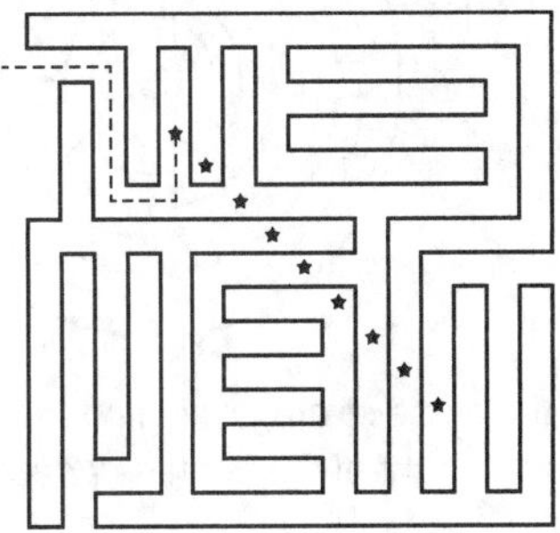

We can reach the next star below by crossing one line. Crossing a line takes us from outside the shape to inside the shape. So, the next star below is inside the shape. We circle it as shown.

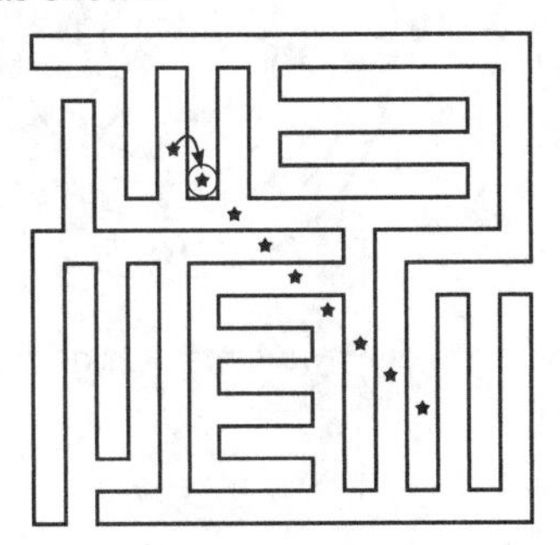

Similarly, we cross one line to get from each star below to the next. Every time we cross a line, we switch from inside to outside, or from outside to inside. So, every other star as we move diagonally to the right is inside the shape. We circle these stars as shown.

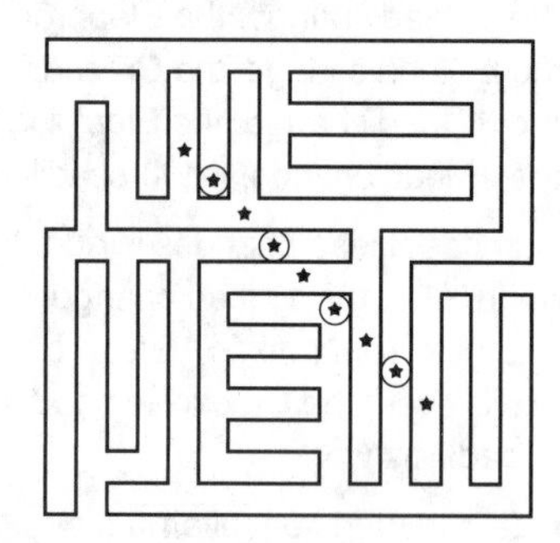

91. We circle the bottom-most star that is outside the shape.

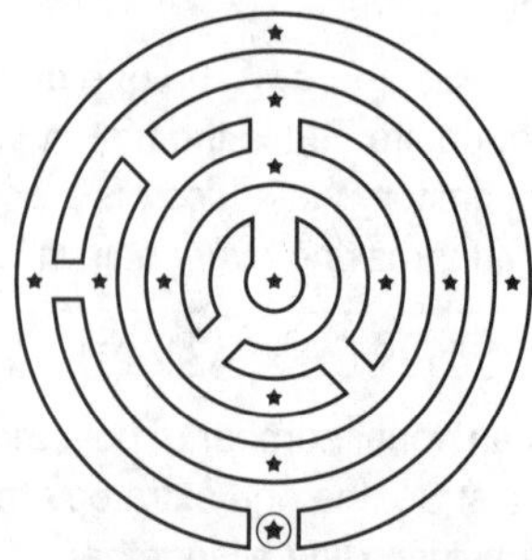

Then, we trace a path from this circled star to the stars directly above it. Every time we cross a line, we go from outside to inside, or from inside to outside.

Keeping track of whether we are inside or outside each time we cross a line, we see that only the center star along this path is also outside the shape.

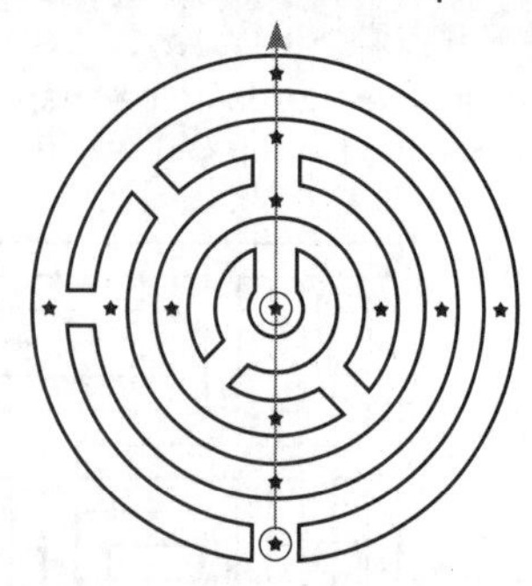

Using the same strategy, we find that all of the stars to the left and right of the center star are inside the shape.

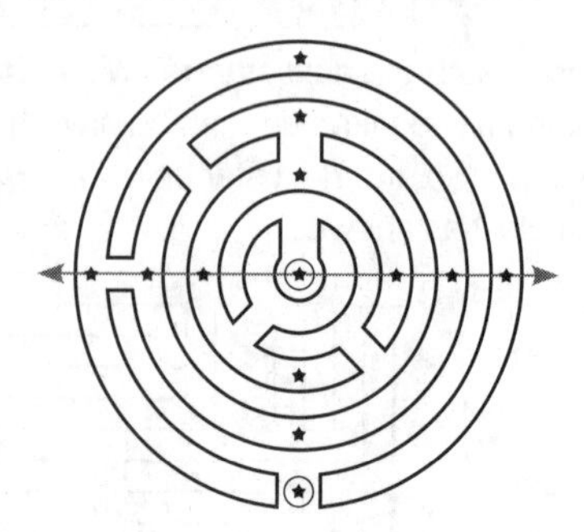

So, there are only two stars outside the shape.

92. We draw a path from the outside of the shape to each letter, as shown below. You may have drawn different paths.

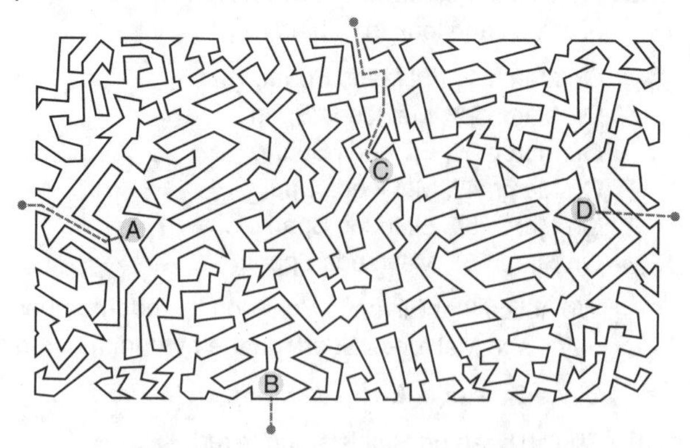

To get to A from the outside, we cross 2 lines, which takes us inside then outside. So, A is **outside** the shape.

To get to B from the outside, we cross 1 line, which takes us inside. So, B is **inside** the shape.

To get to C from the outside, we don't cross any lines. So, C is **outside** the shape.

To get to D from the outside, we cross 3 lines, which takes us inside, then outside, then inside. So, D is **inside** the shape.

93. In the previous solution, we showed one possible path Billy Bug could take to get to each letter from the outside of the shape.

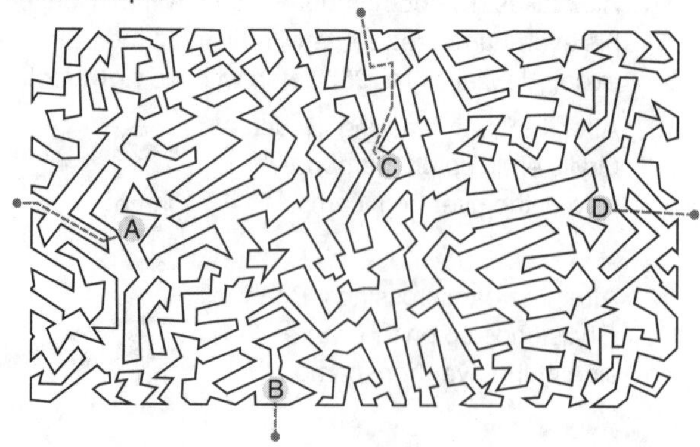

To get to A, Billy Bug can cross 2 lines, and 2 is **even**.
To get to B, Billy Bug can cross 1 line, and 1 is **odd**.
To get to C, Billy Bug can cross 0 lines, and 0 is **even**.
To get to D, Billy Bug can cross 3 lines, and 3 is **odd**.

— or —

Each time Billy Bug crosses a line, he switches from outside to inside, or from inside to outside.

Billy Bug starts outside the shape. To reach a letter that is inside the shape, he must cross an odd number of lines. To reach a letter that is outside the shape, he must cross an even number of lines.

In the previous problem, we learned the following:

- A is outside the shape.
- B is inside the shape.
- C is outside the shape.
- D is inside the shape.

Since A is outside the shape, Billy Bug must cross an **even** number of lines to reach A.

Since B is inside the shape, Billy Bug must cross an **odd** number of lines to reach B.

Since C is outside the shape, Billy Bug must cross an **even** number of lines to reach C.

Since D is inside the shape, Billy Bug must cross an **odd** number of lines to reach D.

94. We draw a path Bobby Bug can take between each given pair of letters.

From A ➡ B, Bobby Bug can cross 9 lines, and 9 is **odd**.
From A ➡ C, Bobby Bug can cross 8 lines, and 8 is **even**.
From B ➡ C, Bobby Bug can cross 7 lines, and 7 is **odd**.
From B ➡ D, Bobby Bug can cross 8 lines, and 8 is **even**.

— *or* —

Bobby Bug ends up on the *same* side of the shape he starts on if he crosses an *even* number of lines. He ends up on the *opposite* side of the shape he starts on if he crosses an *odd* number of lines.

In the previous problem, we learned the following:

- A is outside the shape.
- B is inside the shape.
- C is outside the shape.
- D is inside the shape.

From A ➡ B, Bobby Bug goes from outside to inside. So, he crosses an **odd** number of lines.

From A ➡ C, Bobby Bug goes from outside to outside. So, he crosses an **even** number of lines.

From B ➡ C, Bobby Bug goes from inside to outside. So, he crosses an **odd** number of lines.

From B ➡ D, Bobby Bug goes from inside to inside. So, he crosses an **even** number of lines.

ODDS & EVENS

Coins 92–93

95. Each move turns over *two* coins. We can turn all coins to tails with the moves below.

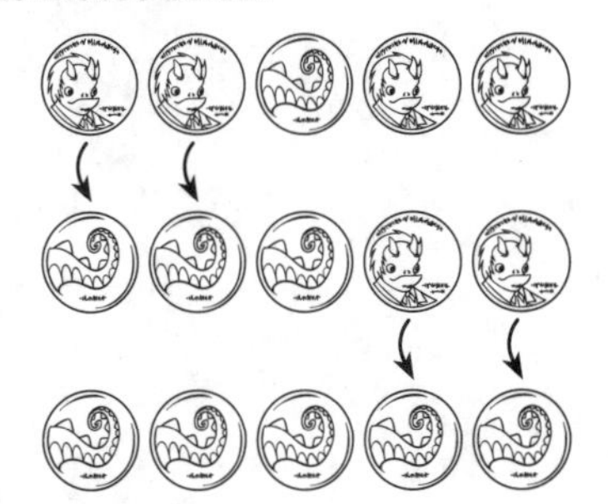

It is not possible to turn all coins to tails in 1 move. So, **2** is the fewest moves we can make to turn all coins to tails.

96. Every time we turn over two coins, one of the following occurs:

- We turn 2 heads to tails.
  This decreases the number of heads by 2.
- We turn 2 tails to heads.
  This increases the number of heads by 2.
- We turn 1 heads and 1 tails to 1 tails and 1 heads.
  This does not change the number of heads.

So, the number of heads can only go up or down by 2. We start with 5 heads, which is odd. Changing an odd by 2 always gives an odd result. So, we can never get an even number of heads.

Getting 0 heads (or 5 tails) is **impossible**.

97. There are 5 heads and 2 tails. In the previous problem, we learned that we can only change the number of heads by 2. Since 5 is odd, we can never get an even number of heads.

So, getting 0 heads (or 7 tails) is **impossible**.

98. We can turn all eight coins to tails with the moves below.

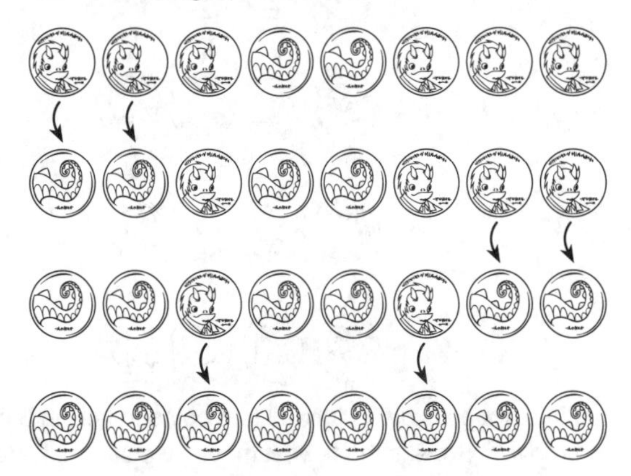

It is not possible to turn all eight coins to tails in 2 or fewer moves. So, **3** is the fewest moves we can make to turn all eight coins to tails.

99. Each move turns over *three* coins. We can turn all six coins to tails with the moves below.

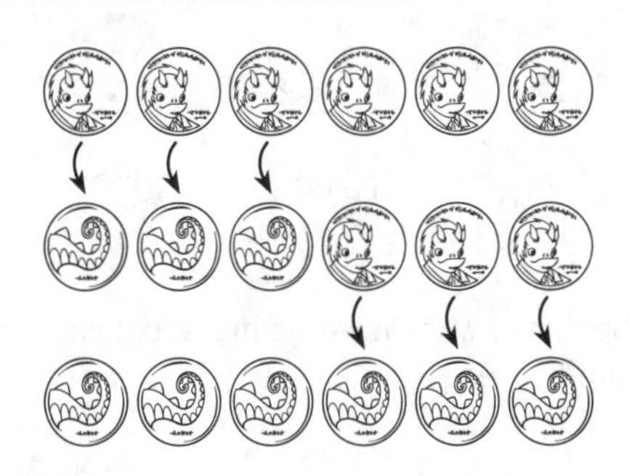

It is not possible to turn all six coins to tails in 1 move. So, **2** is the fewest moves we can make to turn all six coins to tails.

100. We can turn all five coins to tails with the moves below.

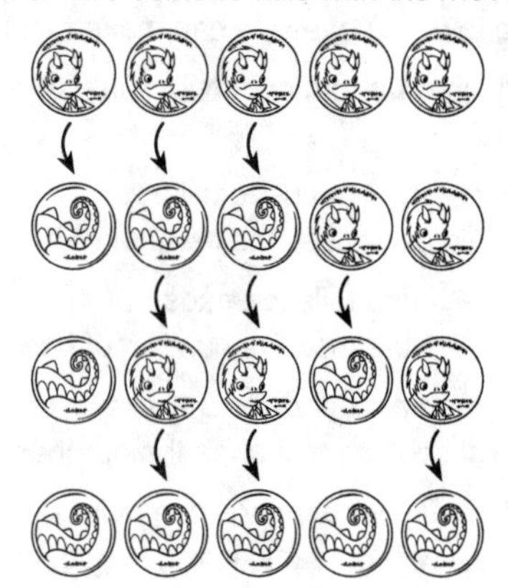

It is not possible to turn all five coins to tails in 2 or fewer moves. So, **3** is the fewest moves we can make to turn all five coins to tails.

101. We can turn all four coins to tails with the moves below.

It is not possible to turn all four coins to tails in 3 or fewer moves. So, **4** is the fewest moves we can make to turn all four coins to tails.

102. There are four coins all together. Since we must turn over three coins that are next to each other, there are only two possible moves: turning over the left three coins, or turning over the right three coins.

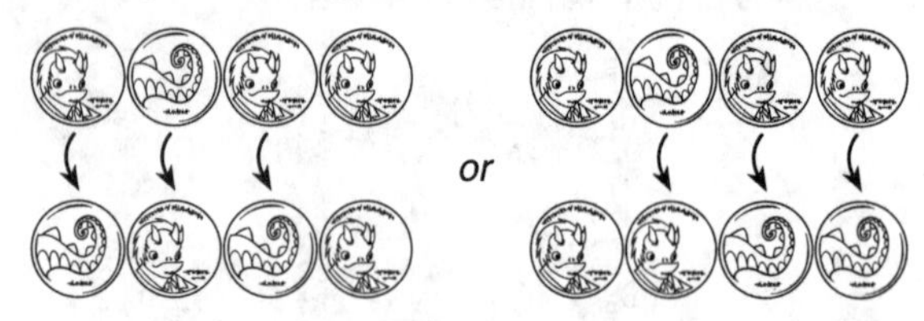

No matter what move we make, both middle coins will be turned over.

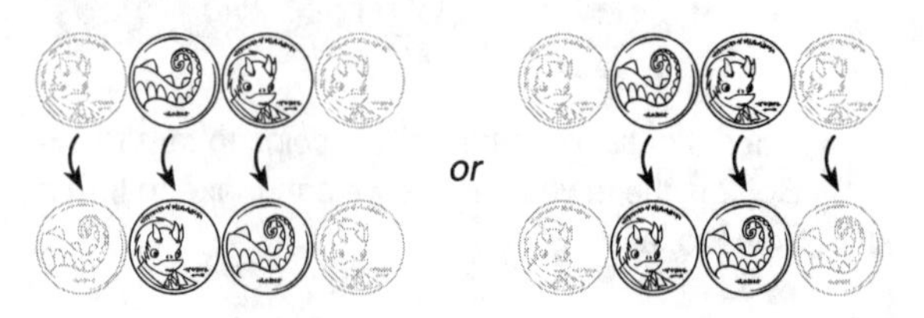

So, the two middle coins will always have opposite sides showing. Since it is impossible for both middle coins to be tails, it is **impossible** for all four coins to be tails.

103. We can turn all five coins to tails with the moves below.

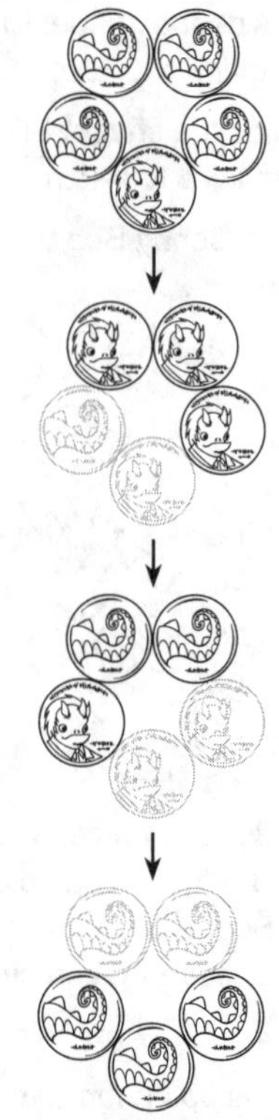

It is not possible to turn all five coins to tails in 2 or fewer moves. So, **3** is the fewest moves we can make to turn all five coins to tails.

ODDS & EVENS Checkerboard Paths 94–95

104. There are six different ways to trace a path from Start to Finish that passes through each square once. We show all six possibilities below.

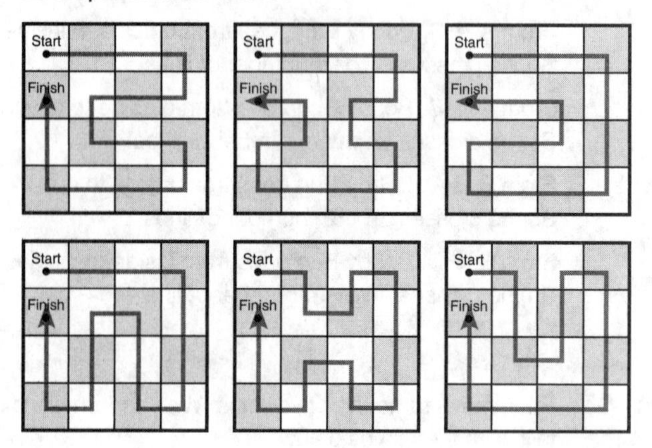

105. There are eight different ways to trace a path from Start to Finish that passes through each square once. We show all eight possibilities below.

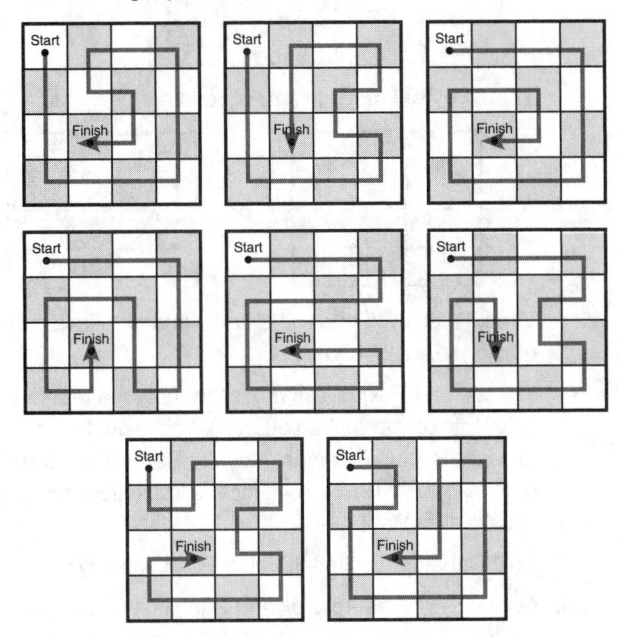

106. No matter what paths we try, we cannot end on the Finish square without leaving at least one square untouched.

Below are just a few paths that do not work.

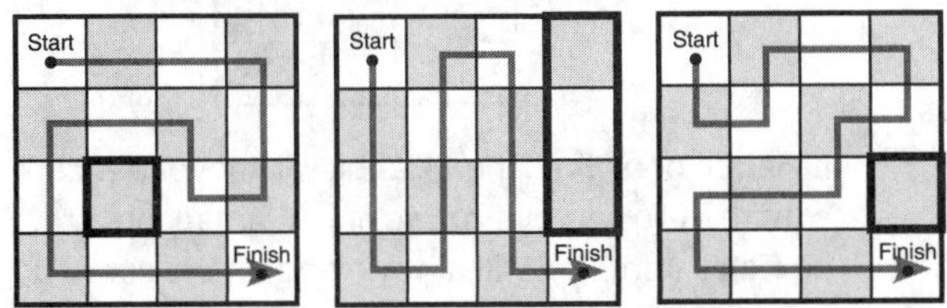

It is **impossible**, so we circle the checkerboard.

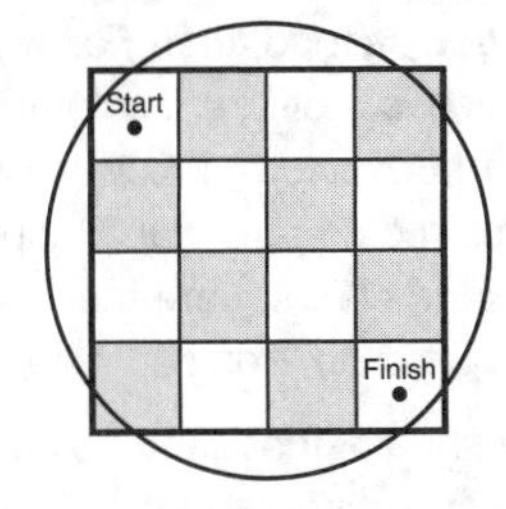

We explore why this is impossible in the problems that follow.

107. Below, we number the squares so that a path can be traced through the squares in order from 1 to 16.

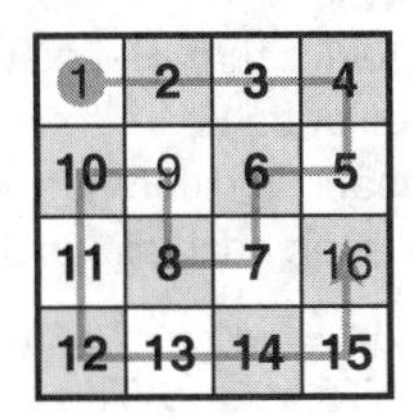

This is the only solution.

108. It is **impossible** to number the squares so that a path can be traced through the squares in order from 1 to 16. So, we circle the checkerboard.

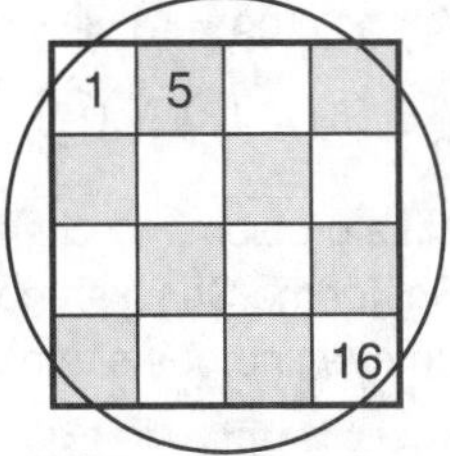

We explore why this is impossible in the problems that follow.

109. Below, we number the squares so that a path can be traced through the squares in order from 1 to 16.

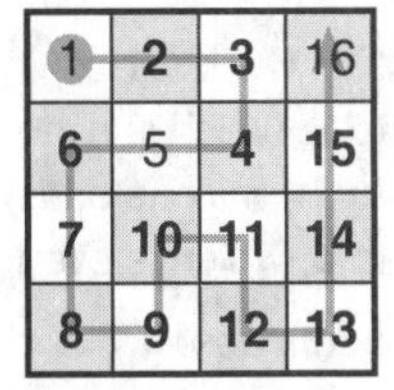

This is the only solution.

110. Below, we number the squares so that a path can be traced through the squares in order from 1 to 16.

1	2	3	4
14	15	6	5
13	16	7	8
12	11	10	9

This is the only solution.

111. Below, we number the squares so that a path can be traced through the squares in order from 1 to 16.

1	2	11	12
4	3	10	13
5	8	9	14
6	7	16	15

This is the only solution.

112. It is **impossible** to number the squares so that a path can be traced through the squares in order from 1 to 16. So, we circle the checkerboard.

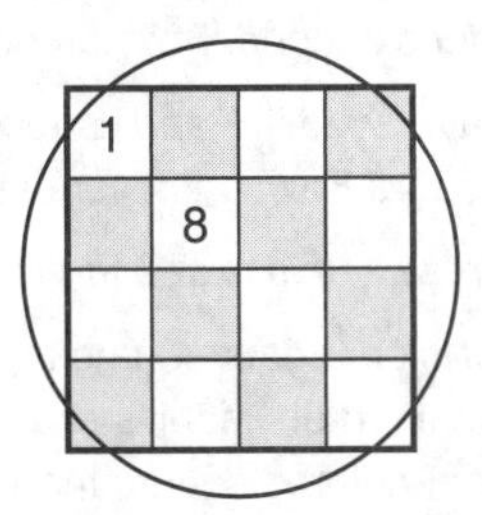

We explore why this is impossible in the problems that follow.

113. The checkerboards in Problems 107, 109, 110, and 111 are possible to number.

We circle all of the numbers that appear in dark squares in Problems 107, 109, 110, and 111.

1 (2) 3 (4) 5 (6) 7 (8) 9 (10) 11 (12) 13 (14) 15 (16)

114. In the previous problem, we circled all of the even numbers. So, for the Checkerboard Paths that are possible, the even numbers appear in dark squares.

(dark squares) light squares dark and light squares

115. In each Checkerboard Path, we must always go from a light square to a dark square, or from a dark square to a light square.

We start with 1, which is odd, in a light square.
Then we go to 2, which is even, in a dark square.
Then we go to 3, which is odd, in a light square.
Then we go to 4, which is even, in a dark square.

This pattern continues. If 1 is in a light square, all odd numbers must be in light squares, and all even numbers must be in dark squares. If this is not true, then the Checkerboard Path is impossible!

The 2nd, 4th, and 5th boards below have an even number in a light square. So, they are impossible to number.

The 6th board below has an odd number in a dark square. So, it is impossible to number.

We circle these impossible boards as shown.

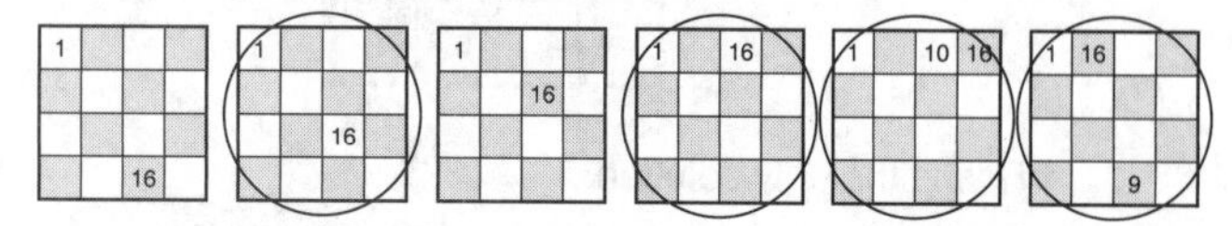

Check: We make sure the 1st and 3rd boards can be numbered, as shown below.

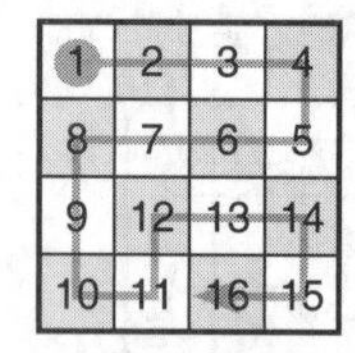

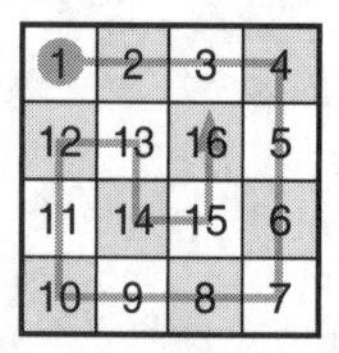

Look again at Problems 106, 108, and 112. Can you use the reasoning above to show why they are impossible?

116. If Rabbit goes first, Fox can eventually catch Rabbit. If Rabbit goes second, Fox can never catch Rabbit.

So, the player who is Rabbit should choose to go **second** so that Fox can never catch Rabbit.

We explore why this is true in the problems that follow.

117. Rabbit starts on a dark square and Fox starts on a light square. On each of Rabbit's and Fox's turns, they move to a different-colored square than the one they were on.

So, on the first turn Rabbit will move to a light square.

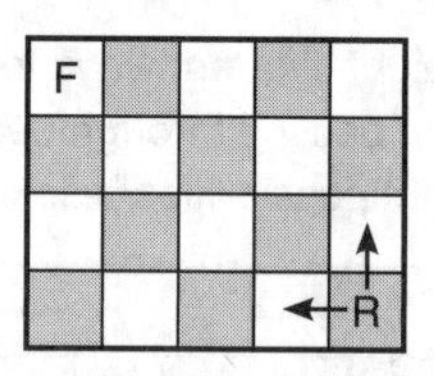

Then, Fox will move to a dark square.

or

Then, Rabbit will move to a dark square, followed by Fox moving to a light square.

We see that Fox always moves to a square that is a different color than the square Rabbit is on. But, for Fox to catch Rabbit, Fox must move to Rabbit's square. This can't happen if Fox always moves to a different-colored square than Rabbit.

So if Rabbit goes first, then **no**, Fox cannot win.

118. On the 5-by-5 board, Fox and Rabbit both start on light squares.

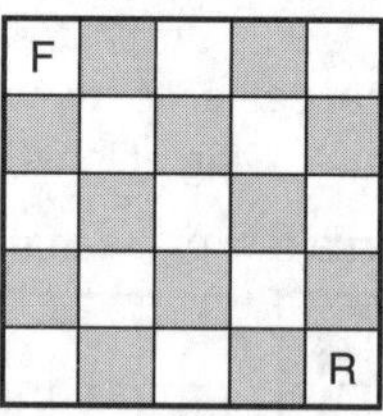

If Rabbit goes first, then Rabbit will move to a dark square, followed by Fox moving to a dark square. So, on Fox's turn, Fox will always land on the same-colored square as Rabbit's square. This means that if Fox plays well, Fox can corner Rabbit and land on Rabbit's square.

If Rabbit goes second, then Fox will move to a dark square, which is a different color than the square Rabbit is on. So, on Fox's turn, Fox will always land on a different-colored square than Rabbit's square. Since Fox can never land on a square that is the same color as the square Rabbit is on, Fox can never catch Rabbit.

So, Rabbit should choose to go **second** on a 5-by-5 board to avoid being caught.

119. We can color any Fox & Rabbit board like a checkerboard. Squares that are next to each other are always different colors. Since Fox is in a square next to Rabbit, Fox and Rabbit are on different-colored squares.

Since it is Rabbit's turn, Rabbit will move to a square that is the same color as Fox's square. Then, Fox will move to a square that is a different color than Rabbit's square. This will repeat on each player's turn.

So, Fox can never land on a square that is the same color as Rabbit's square. This means that Fox can never land on Rabbit's square.

So, **Rabbit** will win the game.

Can you use what you have learned to explain why Rabbit should go second in Problem 116?

120. There are several ways to trace every gray line without picking up our pencil or tracing the same line twice. One way is shown below.

121. There are many ways to trace every gray line without picking up our pencil or tracing the same line twice. One way is shown below.

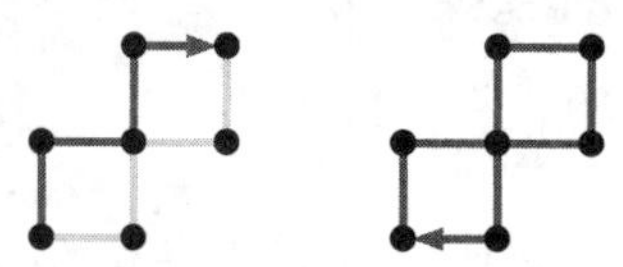

122. This group of dots is impossible to trace. So, we circle it.

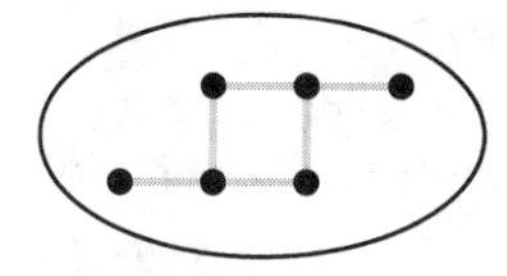

We explore why this is impossible in problems 129-132.

123. There are many ways to trace every gray line without picking up our pencil or tracing the same line twice. One way is shown below.

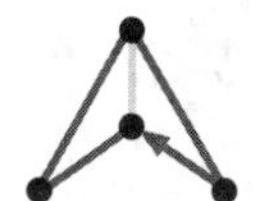

124. This group of dots is impossible to trace. So, we circle it.

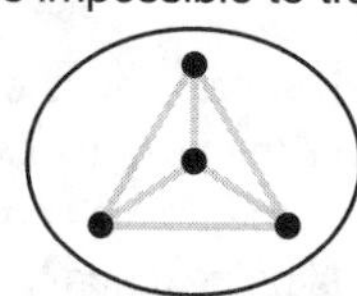

We explore why this is impossible in problems 129-132.

125. There are many ways to trace every gray line without picking up our pencil or tracing the same line twice. One way is shown below.

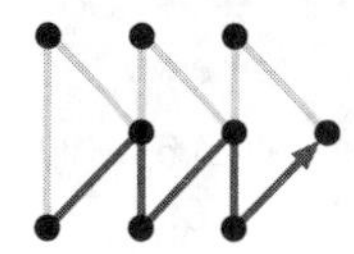
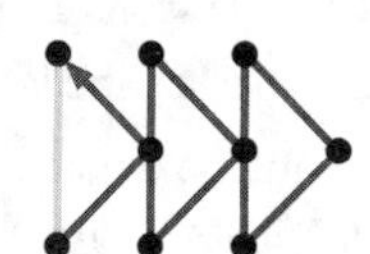
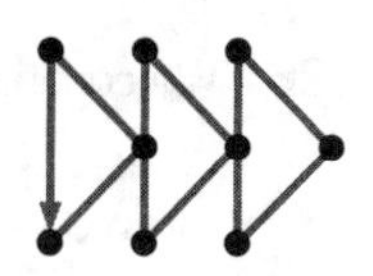

126. This group of dots is impossible to trace. So, we circle it.

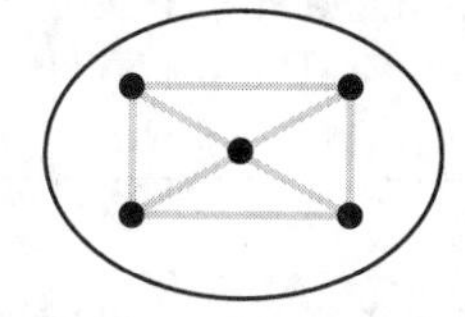

We explore why this is impossible in problems 129-132.

127. There are many ways to trace every gray line without picking up our pencil or tracing the same line twice. One way is shown below.

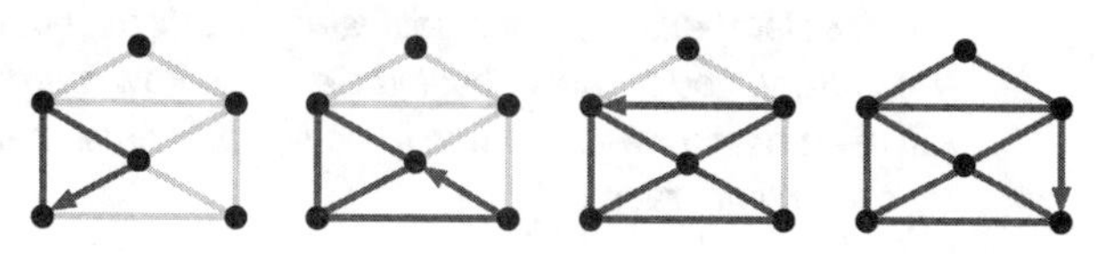

128. There are many ways to trace every gray line without picking up our pencil or tracing the same line twice. One way is shown below.

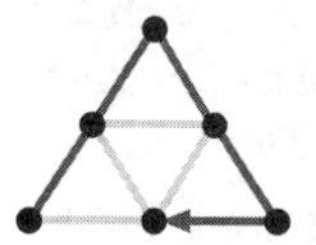
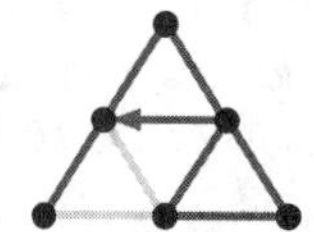
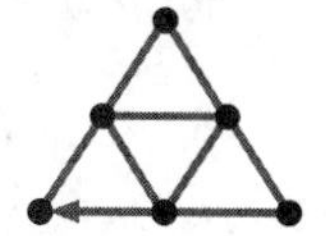

129. We circle all of the odd dots in Problems 120-128 as shown below.

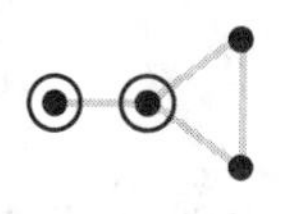
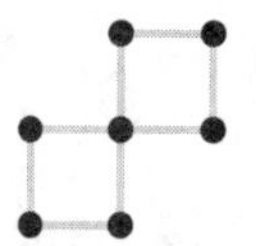
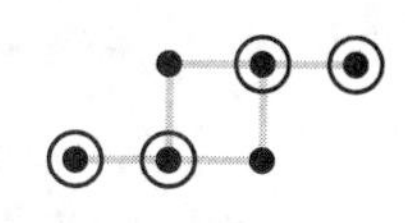

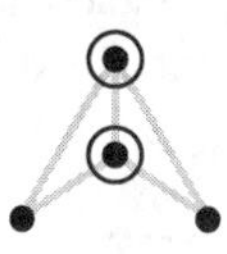
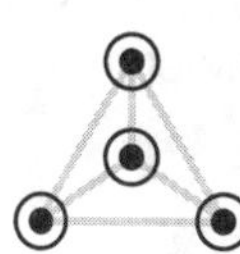
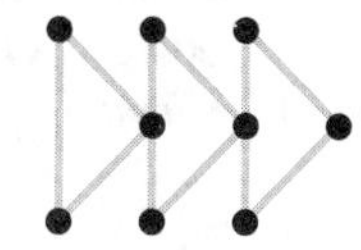

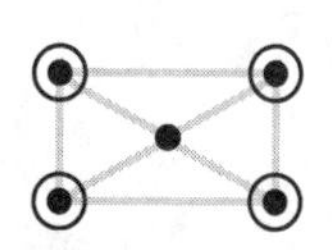
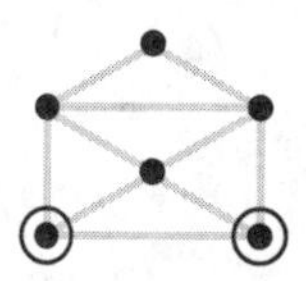
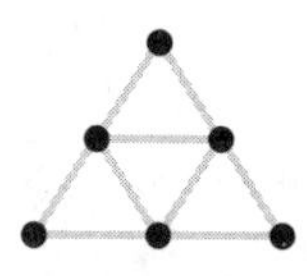

130. The dots are all even in Problems 121, 125, and 128. These problems are all possible to trace. So, if all the dots are even, it is **possible** to trace every line.

There are exactly 2 odd dots in Problems 120, 123, and 127. These problems are all possible to trace. So, if there are exactly 2 odd dots, it is **possible** to trace every line.

There are more than 2 odd dots in Problems 122, 124, and 126. These problems are all impossible to trace. So, if there are more than 2 odd dots, it is **impossible** to trace every line.

This isn't just true for Problems 120-128. It's true for any connected group of dots!

If a dot is in the middle of a path, every path "in" must be followed by a path "out." So, dots in the middle of the path have the same number of "ins" and "outs".

This means that all dots in the middle of a path are even.

So, an odd dot can only be the start or end of a path.

There is only 1 start and 1 end, so it is impossible to have more than 2 odd dots.

Showing why it's ***always*** *possible to trace groups with all even dots or exactly 2 odd dots is more difficult, but we encourage you to think about it! Can you draw a group of all even dots, or a group with exactly 2 odd dots, that is impossible to trace?*

131. Problems 120, 123, and 127 have exactly 2 odd dots. To trace these problems, we must start and end on an <u>**odd**</u> dot.

This is true for any connected group of dots with exactly 2 odd dots! In the solution to Problem 130, we showed that an odd dot can only be the start or end of a path. So, if there are 2 odd dots, one must be the start of the path and the other must be the end of the path.

132. We circle the odd dots in each drawing as shown below.

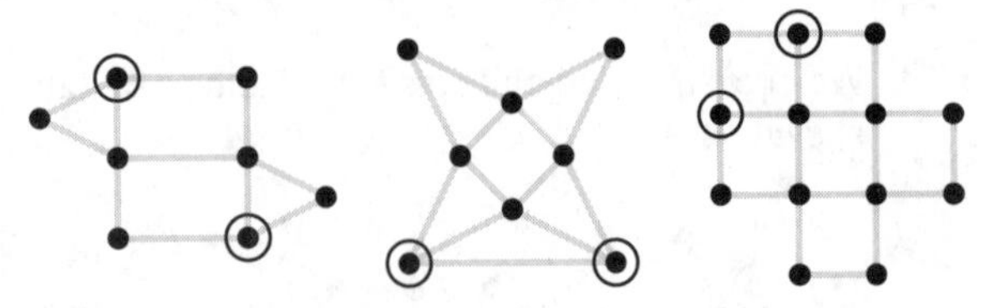

We see that each drawing has exactly 2 odd dots. In the previous problem, we saw that we can only trace drawings with 2 odd dots by starting and ending on an odd dot. We use this strategy to trace each drawing as shown.

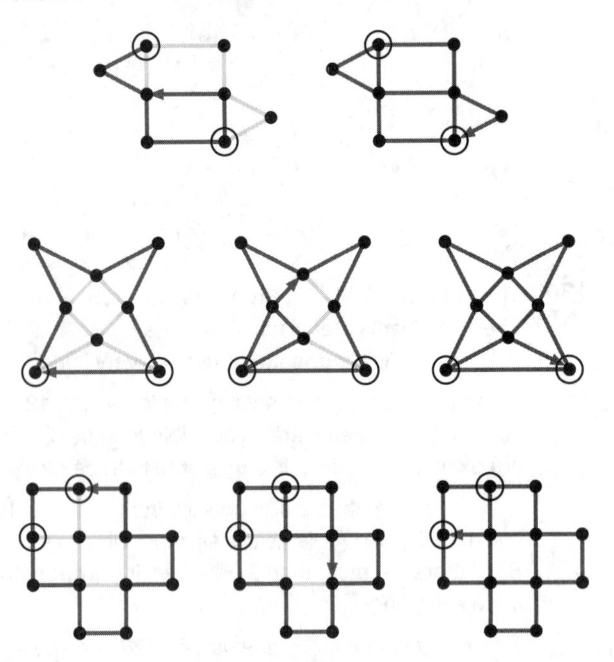

There are many other ways to trace each drawing, all of which start and end on an odd dot!

133. Consecutive numbers alternate between even and odd. So, for any four consecutive numbers, exactly 2 will be odd and 2 will be even. The sum of 2 odds and 2 evens is even. So, Rosa's sum must be even.

Only 44 and 66 are even. We guess-and-check to see if we can find four consecutive numbers that sum to 44 or 66.

$9+10+11+12=42$ and $10+11+12+13=46$. So, there are no four consecutive numbers that sum to 44.

But, $15+16+17+18=66$. So, 66 is the only number that could be Rosa's result.

33 44 55 (66) 77

134. Drawing two lines will split the dots into either three or four groups. Each group must have an odd number of dots. Since the sum of three odds is odd, there cannot be three groups of dots. So, there must be four groups.

Each group must have a different number of dots. There is only one way to add four different odds to get 16:

$$1+3+5+7=16.$$

So, the two lines must separate the dots into groups of 1, 3, 5, and 7. There are several ways to do this. We show a few different ways below.

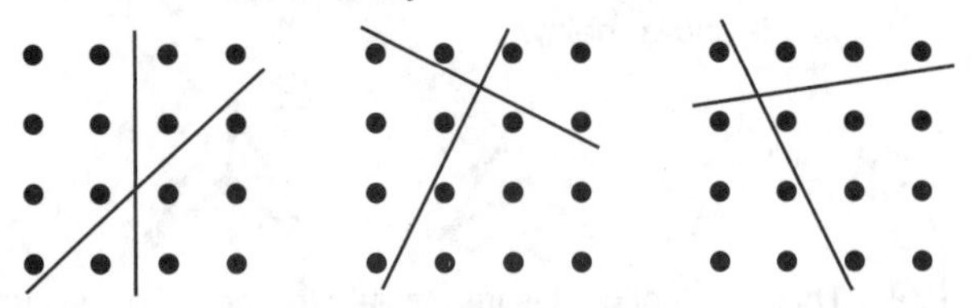

135. Since 57 is odd, the next number is $57+3=60$.
Since 60 is even, the next number is half of 60, which is 30.
Since 30 is even, the next number is half of 30, which is 15.
Since 15 is odd, the next number is $15+3=18$.
Since 18 is even, the next number is half of 18, which is 9.
Since 9 is odd, the next number is $9+3=12$.
Since 12 is even, the next number is half of 12, which is 6.

So, we fill the blanks in Lizzie's list as shown below.

57, <u>**60**</u>, <u>**30**</u>, <u>**15**</u>, <u>**18**</u>, <u>**9**</u>, <u>**12**</u>, <u>**6**</u>

136. We continue Lizzie's list and look for patterns.

Since 6 is even, the next number is half of 6, which is 3.
Since 3 is odd, the next number is $3+3=6$.
Since 6 is even, the next number is half of 6, which is 3.
Since 3 is odd, the next number is $3+3=6$.

We notice that after 12, Lizzie's numbers alternate between 6 and 3. This continues forever!

1st	2nd	3rd	4th	5th	6th	7th	8th	9th	10th	11th	12th	13th
57	60	30	15	18	9	12	6	3	6	3	6	3

After 12, all of the "eventh" numbers are 6, and all of the "oddth" numbers are 3. Since 100 is even, the 100th number in Lizzie's list is **6**.

137. Mike splits the original pile of pennies into two equal piles. This is only possible if the number of pennies is even. The original pile had between 30 and 40 pennies. So, it must have had 32, 34, 36, or 38 pennies.

If we split each of the above amounts into two equal piles, we get the following results.

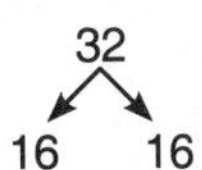

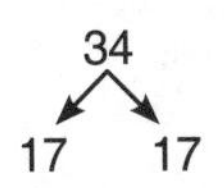

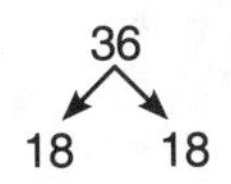

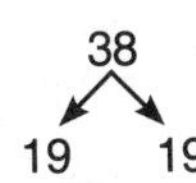

Mike splits each of the two smaller piles into two equal piles. This is only possible if the smaller piles have an even number of pennies. So, this leaves two possibilities.

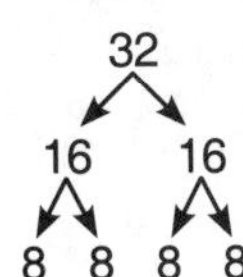

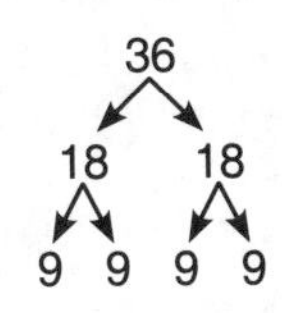

Mike can no longer separate each pile into two equal piles. So, each pile has an odd number of pennies.

So, Mike must have finished with 9 pennies in each pile, which means that he started with **36** pennies.

138. Using the strategies we learned in Problems 120-132, we begin by circling the odd dots in the drawing.

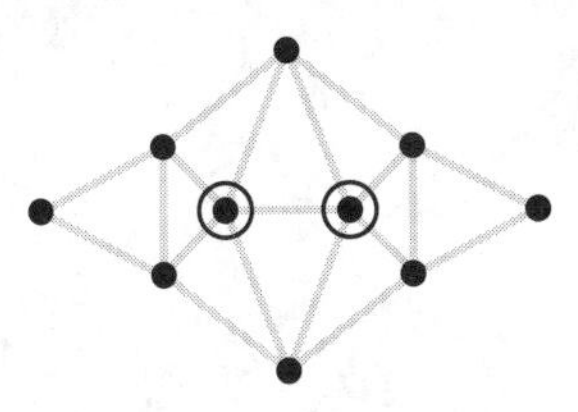

There are exactly 2 odd dots. So, we trace a path that starts and ends on an odd dot. One way to do this is shown below.

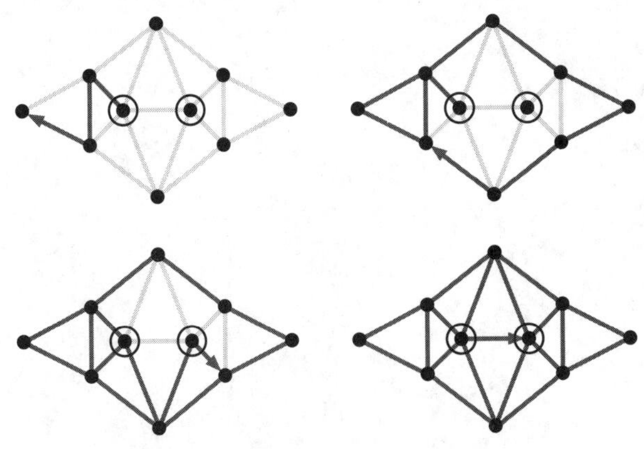

There are many other ways to trace this drawing, all of which start and end on one of the two odd dots!

139. The 5th number in Winnie's list is either odd or even. We consider both possibilities.

- If the 5th number is odd, then the 7th and 9th numbers are also odd, and the 6th and 8th numbers are even. So, every statement except for the second one is true.

If Winnie's 5th number is odd:

Winnie's 5th number is odd.	True
Winnie's 6th number is odd.	False
Winnie's 7th number is odd.	True
Winnie's 8th number is even.	True
Winnie's 9th number is odd.	True

- If the 5th number is even, then the 7th and 9th numbers are also even, and the 6th and 8th numbers are odd. So, only the second statement is true.

If Winnie's 5th number is even:

Winnie's 5th number is odd.	False
Winnie's 6th number is odd.	True
Winnie's 7th number is odd.	False
Winnie's 8th number is even.	False
Winnie's 9th number is odd.	False

We are told that only one statement is true, so it must be the second statement. We circle the second statement as shown.

Winnie's 5th number is odd.

(Winnie's 6th number is odd.)

Winnie's 7th number is odd.

Winnie's 8th number is even.

Winnie's 9th number is odd.

For additional books, printables, and more, visit
BeastAcademy.com